Single-photon Detection for Data Communication and Quantum Systems

IOP Series in Advances in Optics, Photonics and Optoelectronics

SERIES EDITOR

Professor Rajpal S Sirohi Consultant Scientist

About the Editor

Rajpal S Sirohi is currently working as a faculty member in the Department of Physics, Alabama A&M University, Huntsville, Alabama (USA). Prior to this, he was a consultant scientist at the Indian Institute of Science, Bangalore, and before that, he was chair professor in the Department of Physics, Tezpur University, Assam. From 2000 to 2011, he was an academic administrator, being vice chancellor to a couple of universities and the director of the Indian Institute of Technology, Delhi. He is the recipient of many international and national awards and the author of more than 400 papers. Dr Sirohi is involved with research into optical metrology, optical instrumentation, holography, and speckle phenomena.

About the series

Optics, photonics, and optoelectronics are enabling technologies in many branches of science, engineering, medicine and agriculture. These technologies have reshaped our outlook and our ways of interacting with each other and have brought people closer. They help us to understand many phenomena better and provide a deeper insight in the functioning of nature. Further, these technologies themselves are evolving at a rapid rate. Their applications encompass a very large spatial range from the nanometer scale to the astronomical scale and a very large temporal range from picoseconds to billions of years. This series on advances in optics, photonics, and optoelectronics aims to cover topics that are of interest to both academia and industry. Some of the topics that the books in the series will cover include biophotonics and medical imaging, devices, electromagnetics, fiber optics, information storage, instrumentation, light sources, CCD and CMOS imagers, metamaterials, optical metrology, optical networks, photovoltaics, freeform optics and its evaluation, singular optics, cryptography, and sensors.

About IOP ebooks

Authors are encouraged to take advantage of the features made possible by electronic publication to enhance the reader experience through the use of colour, animation, and video, and incorporating supplementary files in their work.

Do you have an idea for a book you'd like to explore?

For further information and details of submitting book proposals see iopscience.org/books or contact Ashley Gasque on Ashley.gasque@iop.org.

Single-photon Detection for Data Communication and Quantum Systems

Michael Hofbauer, Kerstin Schneider-Hornstein and Horst Zimmermann
Technische Universität Wien, Vienna, Austria

IOP Publishing, Bristol, UK

ISBN 978-0-7503-2584-4 (ebook)
ISBN 978-0-7503-2582-0 (print)
ISBN 978-0-7503-2585-1 (myPrint)
ISBN 978-0-7503-2583-7 (mobi)

DOI 10.1088/978-0-7503-2584-4

Version: 20211201

IOP ebooks

British Library Cataloguing-in-Publication Data: A catalogue record for this book is available from the British Library.

Published by IOP Publishing, wholly owned by The Institute of Physics, London

IOP Publishing, Temple Circus, Temple Way, Bristol, BS1 6HG, UK

US Office: IOP Publishing, Inc., 190 North Independence Mall West, Suite 601, Philadelphia, PA 19106, USA

This book is dedicated to our families.

Contents

Preface

The era of photodiode integrated circuits (PDICs) started in the 1990s with their application in audio CD players, CD-ROMs, and DVD systems; this occurred because the bandwidths and transimpedance of optical sensors consisting of discrete photodiodes/photodiode arrays and amplifier integrated circuits (ICs) reached a limit at that time. Since then, rapid progress has been made in the bandwidth/data rate and quantum efficiency/responsivity of integrated photodiodes and sensor ICs. The first quantum leap was the integrated p–intrinsic–n photodiode. Avalanche photodiodes (APDs) in the linear mode then introduced the second step in performance gain. The newest quantum leap in integrated photodetectors took the form of APDs operated in the Geiger mode, in which the detection of single photons is possible.

Integrated photodiodes not only allowed increased bandwidth and transimpedance gains; they were essential for image sensors with high and very high numbers of pixels. Integrated photodiodes offered better immunity to electromagnetic interference, improved reliability, and, last but not least, they reduced the costs of many optical sensors.

Single-photon detection has been the subject of huge publicity in the research field for more than a decade; much progress has been made using single-photon avalanche diodes (SPADs) integrated into complementary metal–oxide–semiconductor (CMOS) chips. Many publications cover the topics of SPADs and SPAD sensor ICs for biomedical applications, imaging, and distance measurement/three-dimensional sensors. Publications on the use of integrated SPADs for quantum communications, quantum cryptography, and quantum computer applications seem to be underrepresented. In addition, there is a new trend toward SPAD-based optical receivers for data and free-space communications. This book will therefore focus on the use of SPAD ICs in data communications and quantum systems.

The authors would like to thank Ashley Gasque from IOP Publishing for proposing the idea for this book. They also wish to thank Dr Bernhard Goll, Dr Hiwa Mahmoudi, Dr Reinhard Enne, Dr Bernhard Steindl, Dr Dinka Milovancev, Dr Alija Dervić, and Saman Kohneh Poushi from our institute for their important and excellent contributions to our work on integrated SPADs. In addition, we would like to thank Wolfgang Einbrodt, Dr Konrad Bach, Detlef Sommer, Dr Alexander Zimmer, and Dr Daniel Gäbler from XFAB Semiconductor Foundries for their long cooperation and for enabling huge progress with silicon PDICs.

Michael Hofbauer, Kerstin Schneider-Hornstein and Horst Zimmermann
Vienna, Austria

Author biographies

Michael Hofbauer

Dr Michael Hofbauer received his Dipl.-Ing. degree in Electrical Engineering from TU Wien (Vienna University of Technology) in 2011. He became a research assistant in 2011 and a university assistant in 2016. In 2017, he received a doctoral degree from TU Wien. He finished his doctoral studies *sub auspiciis Praesidentis* (i.e. with the highest possible honors). His main fields of research are optoelectronic integrated circuits, single-photon detectors, integrated photonics, distance measurements, and single-event effects. He has authored and coauthored more than 70 journal and conference contributions.

Kerstin Schneider-Hornstein

Dr Kerstin Schneider-Hornstein received the Dipl.-Ing. degree and the Dr. techn. degree from the Vienna University of Technology, Austria, in 2000 and 2004, respectively. Since 2001 she has worked for the Vienna University of Technology in the Institute of Electrodynamics, Microwave, and Circuit Engineering, Vienna, Austria. Her major fields of interest are optoelectronics, photonic-electronic integration, and integrated circuit design. She is the author of the Springer book 'Highly Sensitive Optical Receivers' and the author or coauthor of more than 65 journal and conference papers.

Horst Zimmermann

Dr Horst Zimmermann received the Diploma in Physics in 1984 from the University of Bayreuth, Germany, and the Dr.-Ing. degree from the University of Erlangen-Nürnberg while working at the Fraunhofer Institute for Integrated Circuits (IIS-B), Erlangen, Germany in 1991. He was then appointed an Alexander-von-Humboldt Research Fellow at Duke University, Durham, NC, working on diffusion in Si, GaAs, and InP until 1992. In 1993, he joined the Chair for Semiconductor Electronics at Kiel University, Kiel, Germany, where he lectured in optoelectronics and worked on optoelectronic integration. Since 2000 he has been a full professor for Electronic Circuit Engineering at the Vienna University of Technology, Vienna, Austria. His main interests are in the design and characterization of analog and nanometer CMOS circuits as well as optoelectronic integrated CMOS and bipolar CMOS circuits, optical wireless communications, single-photon detection, and electronic–photonic integration. He is the author of the Springer books 'Integrated Silicon Optoelectronics' and 'Silicon Optoelectronic Integrated Circuits' as well as the coauthor of 'Highly Sensitive Optical Receivers,' 'Optical Communication over

Plastic Optical Fibers,' 'Analog Filters in Nanometer CMOS,' 'Comparators in Nanometer CMOS Technology,' and 'Optoelectronic Circuits in Nanometer CMOS Technology.' In addition, he is the author and coauthor of more than 550 publications. In 2002 he became a Senior Member of the IEEE. He was the primary guest editor of the November/December 2014 issue of *IEEE J. Selected Topics in Quantum Electronics on Optical Detectors: Technology and Applications.*

Symbols

$a_{i,j}$	Matrix coefficients
A	Matrix
α	Optical absorption coefficient (μm^{-1})
α	Complex coefficient in the qubit
α	Half of the opening angle of the lens (rad)
α_n	Impact ionization coefficient of electrons (cm^{-1})
α_n	Impact ionization coefficient of holes (cm^{-1})
APP	Afterpulsing probability
β	Complex coefficient in the qubit
BER	Bit error ratio
c	Speed of light in a medium ($cm\,s^{-1}$)
c_0	Speed of light in vacuum ($cm\,s^{-1}$)
c_i	Complex coefficients
d	Resolution (m)
D	Diffusion coefficient ($cm^2\,s^{-1}$)
D_n	Diffusion coefficient of electrons ($cm^2\,s^{-1}$)
D_p	Diffusion coefficient of holes ($cm^2\,s^{-1}$)
DCR	Dark count rate (s^{-1})
DR	Data rate ($Mb\,s^{-1}$)
ε	Permittivity
ε_0	Vacuum permittivity
ε_r	Relative permittivity
$\mathbf{E}$	Electric field ($V\,cm^{-1}$)
f	Frequency
h	Planck constant (J s)
$h\nu$	Photon energy (eV)
I	Current (A)
I_{ph}	Photocurrent (A)
j	Current density ($A\,cm^{-2}$)
k_B	Boltzmann's constant (JK^{-1})
k_BT	Thermal energy (eV)
L	Length (μm)
L_n	Diffusion length of electrons (μm)
L_p	Diffusion length of holes (μm)
λ	Wavelength (m)
μ	Mobility ($cm^2\,V^{-1}\,s^{-1}$)
μ_n	Electron mobility ($cm^2\,V^{-1}\,s^{-1}$)
μ_p	Hole mobility ($cm^2\,V^{-1}\,s^{-1}$)
ν	Frequency of light (Hz)
n	Impurity concentration (cm^{-3})
n	Refractive index
NA	Numerical aperture
N_{counts}	Total number of counts
$N_{detphot}$	Number of detected photons
N_{phot}	Number of incident photons
N_A	Acceptor concentration (cm^{-3})
N_D	Donor concentration (cm^{-3})

OCTP	Optical crosstalk probability (%)
p	Density of free holes (cm^{-3})
P_{opt}	Incident optical power (W)
PDP	Photon detection probability
PDE	Photon detection efficiency
Ψ	Potential (V)
$\vert\Psi\rangle$	Qubit
q	Electron charge (As)
Q_{av}	Avalanche charge (As)
ρ	Charge density (As cm^{-3})
t	Time (s)
τ	Lifetime (s)
τ_n	Electron lifetime (s)
τ_p	Hole lifetime (s)
Θ	Angle (°)
Θ_i	Incidence angle (°)
T	Absolute temperature (K)
t_{meas}	Total measurement time (s)
U	Voltage (V)
U_{BE}	Base–emitter voltage (V)
U_D	Built-in voltage (V)
U_{DS}	Drain–source voltage (V)
U_{GS}	Gate–source voltage (V)
U_T	Thermal voltage k_B T/q (V)
U_{th}	Thermal generation/recombination rate (cm^{-3} s^{-1})
U_{Th}	Threshold voltage (V)
v	Carrier velocity (cm s^{-1})
v_s	Saturation velocity (cm s^{-1})
V_{BD}	Breakdown voltage (V)
V_{EX}	Excess bias voltage of SPAD (V)
W	Width of space-charge region (μm)
ω_i	Frequency of the idler photon (s^{-1})
ω_p	Frequency of the pump laser (s^{-1})
ω_s	Frequency of the signal photon (s^{-1})

Chapter 1

Single-photon avalanche diodes (SPADs)

In this chapter, we describe the basics of optical absorption and photogeneration as well as the semiconductor equations, carrier transport by drift and diffusion, the width of the space-charge region, and the capacitance of photodiodes. These are followed by an introduction to impact ionization, multiplication factor, and breakdown voltage in linear-mode avalanche diodes; the Geiger mode and the properties of single-photon avalanche diodes (SPADs) are then explained. Commercial discrete SPADs are presented before the focus moves to SPADs integrated into complementary metal–oxide–semiconductor (CMOS) and bipolar CMOS (BiCMOS) processes—separated into thin and thick SPADs. The temperature behavior and transient avalanche currents of SPADs are discussed. Thick SPADs are compared in PIN photodiode CMOS, in high-voltage (HV) CMOS, and in modulation-doped SPADs. At the end of this chapter, an outstanding model for photon detection probability is presented, which describes the spectral and excess bias voltage dependencies of PIN-photodiode CMOS and HV CMOS SPADs perfectly. We also present the stunning result that the Lambert–Beer law is not exact close to the silicon surface.

1.1 Basics and properties

Optical absorption and photogeneration

If a photon has an energy that is larger than the bandgap energy of silicon (about 1.12 eV), the photon can be absorbed and an electron–hole pair can be generated in the silicon. The optical power P decays exponentially in matter, following the Lambert–Beer law[1]:

[1] This dependence holds for light that is incident from an infinitely thick medium onto the semiconductor. For light incidence through a thin cover layer into silicon, the optical power deviates from the exponential decay, as we will see in section 1.4.

doi:10.1088/978-0-7503-2584-4ch1

$$P(\lambda, y) = P_0 e^{-\alpha(\lambda)y}, \tag{1.1}$$

where P_0 is the optical power at the silicon surface (y=0), $\alpha(\lambda)$ is the dependence of the optical absorption coefficient on the wavelength λ, and y is the depth (assuming that the incident light is perpendicular to the surface). Among these, $\alpha(\lambda)$ is the most important optical constant for photodetectors; it determines $1/e$, the penetration depth of light inside semiconductors. Table 1.1 lists the optical absorption coefficient of silicon and the $1/e$ penetration depth for some important wavelengths [1, 2], and figure 1.1 depicts the corresponding decay of optical power in silicon.

The dependence of the photogeneration rate per volume G on the depth y in silicon can be obtained using:

$$G(y) = \frac{P(y) - P(y + \Delta y)}{\Delta y} \frac{1}{Ah\nu}, \tag{1.2}$$

Table 1.1. Absorption coefficients α of silicon and the $1/e$ penetration depth for several important wavelengths.

Wavelength (nm)	α (μm^{-1})	$1/e$ penetration depth (μm)
850	0.06	16.67
780	0.12	8.33
680	0.24	4.16
635	0.38	2.63
565	0.73	1.37
465	3.6	0.278

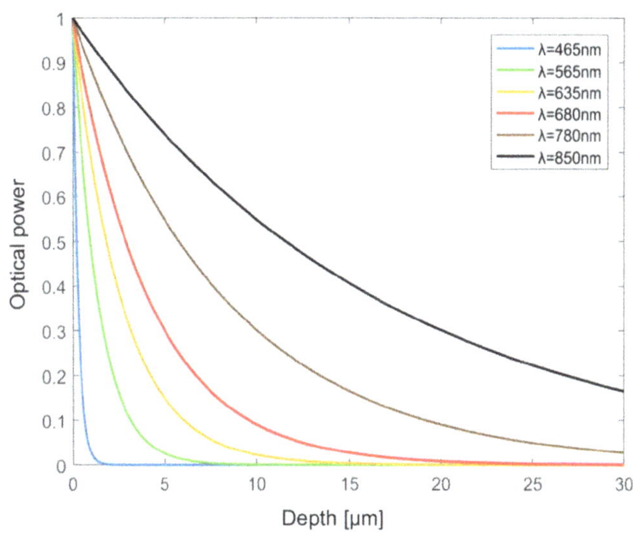

Figure 1.1. Decay of optical power in silicon versus depth (normalized to the optical power at the silicon surface).

where A is the cross-sectional area exposed to incident light and $h\nu$ is the energy $E_{\mathrm{ph}} = \frac{hc}{\lambda}$ of the photon (h = Planck's constant, c = light velocity[2]). For $\Delta y \to 0$, we obtain $G(y) = (-\mathrm{d}P(y)/\mathrm{d}y)\,1/Ah\nu$. After equation (1.1) is differentiated, the following expression for photogeneration is obtained:

$$G(y) = \frac{\alpha P_0}{Ah\nu} e^{(-\alpha y)}. \tag{1.3}$$

Photogeneration has to be considered in the semiconductor equations [3, 4] in order to describe the photocurrent in photodetectors. The interested reader is referred to [5], where these equations can be studied and models of optoelectronic semiconductor devices can be found. Poisson's equation (1.4), the transport equations (equations (1.6) and (1.7)), and the continuity equations (equations (1.9) and (1.10)) collectively constitute the semiconductor equations.

$$\Delta\Psi = -\frac{\rho}{\epsilon}, \tag{1.4}$$

where Ψ is the potential, ρ represents the charge density and ϵ is the product of the relative and absolute dielectric constants: $\epsilon = \epsilon_{\mathrm{r}}\epsilon_0$. The charge density is proportional to the elementary charge (electron charge) q and the sum of the hole concentration p (positively charged), the electron concentration n (negatively charged), the donor concentration N_{D} (positively charged), and the acceptor concentration N_{A} (negatively charged):

$$\rho = q(p - n + N_{\mathrm{D}} - N_{\mathrm{A}}). \tag{1.5}$$

The transport equations, which define the current densities for electrons and holes, are the sum of the drift and diffusion current densities:

$$\vec{j}_{\mathrm{n}} = qn\mu_{\mathrm{n}}\vec{E} + qD_{\mathrm{n}}\,\mathbf{grad}\,n, \tag{1.6}$$

$$\vec{j}_{\mathrm{p}} = qp\mu_{\mathrm{p}}\vec{E} - qD_{\mathrm{p}}\,\mathbf{grad}\,p. \tag{1.7}$$

The total current density is the sum of the electron and hole current densities:

$$\vec{j} = \vec{j}_{\mathrm{n}} + \vec{j}_{\mathrm{p}}. \tag{1.8}$$

The continuity equations can be modified to include thermal generation/recombination U_{th} and the photogeneration G[3] due to the penetration of light into the semiconductor, as follows:

$$\frac{\partial n}{\partial t} = \frac{\mathrm{div}\,\vec{j}_{\mathrm{n}}}{q} + U_{\mathrm{th}} + G, \tag{1.9}$$

[2] Be aware that c and λ depend on the medium in which the light is travelling. Only the photon frequency, ν, is constant.

[3] G depends on the location in the semiconductor for which photogeneration is being calculated; in particular, it depends on y, which is the depth in the semiconductor for perpendicular light incidence.

$$\frac{\partial p}{\partial t} = -\frac{\text{div}\,\vec{j}_{\text{p}}}{q} + U_{\text{th}} + G. \tag{1.10}$$

Drift and diffusion

There are two terms in the transport equations. The first term contains the electric field strength, and the second term is determined by diffusion. The electric field follows from (1.4) and obeys:

$$\vec{E} = -grad\ \Psi. \tag{1.11}$$

The electric field is important for the calculation of the drift velocity $\vec{v}$ of photogenerated carriers in photodiodes:

$$\vec{v} = \mu\vec{E}. \tag{1.12}$$

The mobilities of electrons and holes differ strongly, and therefore either μ_n or μ_p has to be used for μ in order to calculate the electron drift velocity or the hole drift velocity, respectively. The mobilities μ_n and μ_p are only constant for low electric field strengths. The drift velocities of electrons and holes in silicon both saturate at 10^7 cm s^{-1} for large values of the electric field. The carrier mobilities also depend on impurities; in particular, they depend on the dopant concentration. The carrier mobilities decrease with increasing dopant concentration. These factors are important for photodetector development [5]. Device simulators allow the choice of different models for the carrier mobilities, for generation/recombination, and for other quantities [4, 6].

The diffusion of minority carriers is also an important factor in the response speed of photodetectors. Carrier diffusion occurs in semiconductor regions without an electric field. The carrier diffusion coefficients D_n and D_p for electrons and holes, respectively, are determined by the carrier mobilities via the Einstein relation:

$$D_{\text{n/p}} = \mu_{\text{n/p}}\frac{k_{\text{B}}T}{q} = \mu_{\text{n/p}}U_{\text{T}}. \tag{1.13}$$

Therefore, carrier diffusion is usually much slower than carrier drift, because the thermal voltage $U_T = k_B T/q$ (k_B is the Boltzmann constant and T is the absolute temperature) has a low value (26 mV) at around room temperature, but the electric field strength at p/n junctions and in the depleted intrinsic region of p–intrinsic–n (PIN) photodiodes is several thousand V cm^{-1} [5]. Therefore, if light photogenerates carriers only in depleted drift regions, photodiodes exhibit high bandwidth and short photocurrent rise and fall times. If carriers are also photogenerated below the depleted space-charge region in the substrate and in highly doped regions, the photocurrent shows a slow diffusion tail [5]. Figure 1.2 demonstrates the difference.

Another important aspect is the recombination of photogenerated carriers in regions without an electric field. Because minority carriers diffuse more slowly at higher dopant levels, in $n+$ and $p+$ regions as well as in highly doped substrates

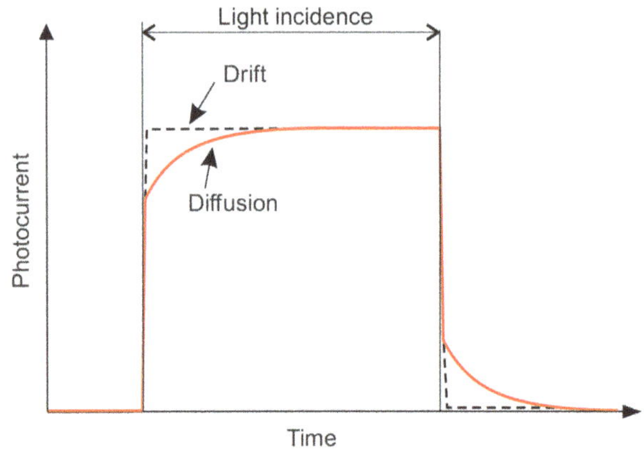

Figure 1.2. Transient response of a photocurrent in the presence of carrier drift and diffusion.

(e. g. in the $p+$ part of $p - /p+$ epitaxial wafers), many photogenerated carriers recombine before they can reach the drift region and therefore do not contribute to the photocurrents of p/n junction photodiodes and PIN photodiodes or to the PDPs of SPADs.

Width of the space-charge region and capacitance

Carrier drift and diffusion are not the only important factors in the time responses of photodetectors and single-photon avalanche diodes. The width of the space-charge region together with the electric field strength in this region determine the drift time and the rise/fall times of the photocurrents of p/n junction and PIN photodiodes. In SPADs, a Geiger-mode event can be delayed after photon absorption by drift through the absorption region. Jitter arises in SPADs because photon absorption can occur at different depths inside the absorption region.

The width of the space-charge region W (and the dependence of the electric field on the location inside the space-charge region) can be obtained from Poisson's equation using the so-called depletion approximation [5]. For an abrupt p/n junction, an analytical solution is obtained:

$$W = \sqrt{\frac{2\epsilon_r\epsilon_0}{q}\frac{N_A + N_D}{N_A N_D}(U_D - U)} \tag{1.14}$$

with

$$U_D = \frac{k_B T}{q}\ln\frac{N_A N_D}{n_i^2}, \tag{1.15}$$

where N_D is the built-in voltage of the p/n junction and n_i is the intrinsic carrier concentration. Because photodiodes work in the reverse direction, a negative value has to be inserted for U in 1.14.

However, the space-charge region width is not only important for the time response of a photodiode. It also determines the capacitance of the photodiode, which has to be considered together with the input resistance of an amplifier circuit to calculate the rise/fall time and the bandwidth of optical sensors or receivers. For SPADs, the capacitance determines the avalanche charge and how fast a quenching or gating circuit can discharge or quench the SPAD. The capacitance of an abrupt p/n junction, in which one side of the p/n junction has several orders of magnitude less dopant than the other, is given by (the lower value of N_A or N_D has to be inserted):

$$C_D = A\sqrt{\frac{q\epsilon_r\epsilon_0 N_{A/D}}{2}}\frac{1}{\sqrt{U_D - U}}, \tag{1.16}$$

Real p/n junctions are not abrupt and so an analytical calculation of W and C is not possible. Fortunately, device simulators such as ATLAS can solve the semiconductor equations [6].

Impact ionisation

In high-electric-field regions within semiconductors, electrons and holes are strongly accelerated and gain high energies, which are sufficient to generate electron–hole pairs by impact ionization. These electron–hole pairs are separated and also accelerated by the high electric field. They can, in turn, also generate electron–hole pairs, and so on. An avalanche is caused and the current grows.

The generation rate G of electron–hole pairs caused by impact ionization is

$$G = \alpha_n n v_n + \alpha_p p v_p, \tag{1.17}$$

where α_n and α_p are the impact ionization coefficients of electrons and holes, n and p are the electron and hole concentrations, and v_n and v_p are the velocities of the electrons and the holes, respectively. The ionization coefficients are measured in units of 1/cm, and $1/\alpha_{n,p}$ can be interpreted as the mean distance travelled between two impact ionization events. G is measured in units of $cm^{-3}\,s^{-1}$. The carrier velocities depend on the electric field [5]. But here, the dependence of the ionization coefficients on the electric field is even more important:

$$\alpha_{n,p}(E) = \frac{qE}{E_{I,n,p}}e^{(-\Xi_{I,n,p}/E)}, \tag{1.18}$$

where E is the electric field strength (in V cm^{-1}) and $E_{I,n,p}$ denotes the high-field, effective ionization threshold energies of electrons and holes, respectively [3]. For electrons in Si, $E_{I,n}$ is 3.6 eV, and for holes in Si, $E_{I,p}$ is 5.0 eV due to ionization scattering events. Ξ_I is the threshold electric field strength at which carriers overcome the decelerating effect of ionization scattering [3]. This critical field strength is about 2×10^5 V cm^{-1} for silicon.

If we assume that a hole current I_{p0} flows from the left-hand side into a high-field region (which starts at $x = 0$) with a width W, the hole current I_p increases with x, and at $x = W$, the hole current $M_p I_{p0}$ leaves the high-field region [3]. The electron current flows in the other direction and increases from W to $x = 0$. In the steady state, the sum I of I_n and I_p (the total current) is constant.

Within an interval dx, the hole current changes (increases) proportionally to the hole current I_p, proportionally to the hole ionization coefficient α_p, and proportionally to the width of the interval, dx. Since high-energy electrons can also produce holes (each carrier generates an electron–hole pair), we have to add a corresponding term for the electrons [3]:

$$dI_p = I_p\alpha_p dx + I_n\alpha_n dx, \tag{1.19}$$

which can be rearranged using $I_n = I - I_p$:

$$\frac{dI_p}{dx} = I_p(\alpha_p - \alpha_n) + \alpha_n I. \tag{1.20}$$

The solution of this equation, considering the boundary condition $I = I_p(W) = M_p I_{p0}$, is [3]:

$$I_p(dx) = \frac{I\left[\frac{1}{M_p} + \int_0^x \alpha_n e^{[-\int_0^x (\alpha_p - \alpha_n)dx']} dx\right]}{e^{[-\int_0^x (\alpha_p - \alpha_n)dx']}}. \tag{1.21}$$

with the hole multiplication factor

$$M_p = \frac{I_p(W)}{I_p(0)}. \tag{1.22}$$

To obtain an expression which has to be fulfilled for avalanche breakdown to take place, equation (1.21) can be rewritten as follows:

$$1 - \frac{1}{M_p} = \int_0^W \alpha_p e^{[-\int_0^x (\alpha_p - \alpha_n)dx']} dx. \tag{1.23}$$

Breakdown voltage

At the avalanche breakdown voltage, the hole multiplication factor approaches infinity. The condition for avalanche breakdown is therefore:

$$\int_0^W \alpha_p e^{[-\int_0^x (\alpha_p - \alpha_n)dx']} dx = 1 \tag{1.24}$$

when holes are injected into the multiplication zone. When electrons initiate the avalanche process, the ionization integral is:

$$\int_0^W \alpha_n e^{[-\int_x^W (\alpha_n - \alpha_p)dx']} dx = 1. \tag{1.25}$$

These equations ((1.24) and (1.25)) are equivalent and also valid for the injection of both electrons and holes. Either equation 1.24 or 1.25 can represent the breakdown condition. So, the breakdown voltage determines the electric field and the electric field determines the ionization coefficients in such a way that the breakdown conditions are fulfilled.

These ionization integrals can be significantly simplified for $\alpha_n = \alpha_p$. However, semiconductor physics and the material properties of Si, GaAs, InGaAs, and other semiconductor materials do not obey this equality.

To give a complete picture, it should be mentioned that the impact ionisation mechanism occurs for Si junction breakdown voltages above $6E_g/q$ (E_g is the bandgap energy and q the electron charge; the E_g of Si is about 1.1 eV at room temperature, so $6E_g/q$ is about 6.6 V). For breakdown voltages below $4E_g/q$, the tunneling effect (band-to-band tunneling) causes a reverse current. Between $4E_g/q$ and $6E_g/q$, the breakdown is due to a mixture of impact ionisation and tunneling [3]. The breakdown voltage depends on the dopant amounts at the p–n junction. For highly doped junctions, the tunneling mechanism dominates, and for lightly doped junctions, the avalanche mechanism prevails. The temperature coefficient of the breakdown voltage is negative for the tunneling mechanism and positive for the avalanche effect. This knowledge can be used to determine which breakdown mechanism dominates in APDs and SPADs.

Geiger mode

Figure 1.3 explains the working principle of SPADs. First, the SPAD is charged to $V_{BD} + V_{EX}$ (the breakdown voltage plus the excess bias voltage); it then waits for photon absorption and the triggering of a Geiger-mode avalanche event. During this waiting phase, the SPAD is at a metastable bias point, which means that no current flows, although the device is biased (far) above the breakdown voltage. When a photon or a thermally generated or tunneling charge carrier finally triggers an avalanche (and this avalanche does not cease), a self-sustaining avalanche builds up and a huge current flows until the voltage drop across a passive quenching resistor reduces the reverse voltage of the SPAD and stops the avalanche, or a circuit detects this avalanche event and switches the SPAD back to its breakdown voltage or less. The current through the passive quenching resistor then charges the SPAD back to $V_{DD} = V_{BD} + V_{EX}$, or a reset circuit charges it to V_{DD} again. It should be mentioned

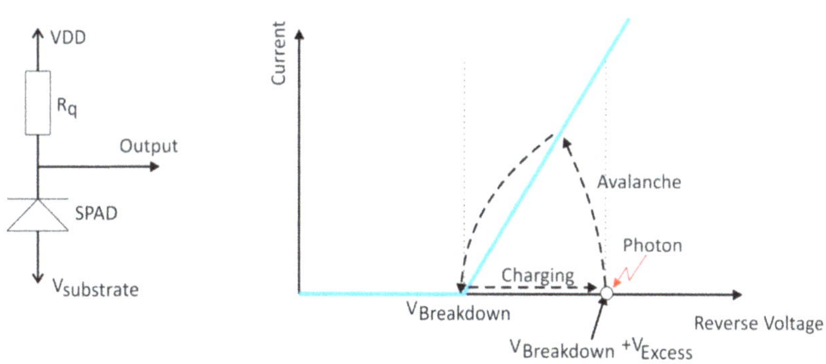

Figure 1.3. Principle of SPAD operation.

that the current curve in figure 1.3 represents steady state-currents for different excess voltages. So, for a larger excess bias, the final avalanche current is larger.

Now, since we have introduced the operational principle of the SPAD in the Geiger mode, we need to answer the question of how the reverse current in the linear mode fits this metastable operation. This question arises since many readers may believe that currents flow continuously in general, and therefore that the reverse current of an avalanche photodiode in the linear mode also flows continuously. However, this is not the case. The fundamental quantum nature of electrical charges, i.e. the electron charge, is the basis for explaining this process. Also, in the linear mode of an APD, i.e. below the breakdown voltage, discrete charge carriers are thermally generated or photogenerated and amplified into charge packages in the multiplication zone. But when the reverse current is measured, the flowing charges or charge packages are averaged over the measurement time. The current is the flowing charge divided by the measurement time. Single-charge packages (in the linear mode, the avalanche gain is usually 10 to 10^3) cannot usually be resolved due to limited time resolution and because the individual charge packages stay below the electronic noise level. In the Geiger mode, however, the amplification in a self-sustaining avalanche can exceed 10^6, and the charge package generated by one photon exceeds the electronic noise level considerably. However, it should also be mentioned that not every photon is detected in the Geiger mode, because, due to the statistical nature of impact ionization, an avalanche can cease before it becomes self-sustaining. Therefore, the photon detection probability of SPADs is often much smaller than 100%. So, an SPAD may detect a single photon, but it will not detect each photon. The theory used to calculate the PDP will be introduced in section 1.4.

We now turn to simple examples of the reverse (dark) current of p–n junctions or PIN photodiodes, in which there is no avalanche gain (i.e. where M=1). For modest junction areas, the reverse current is typically 1 pA, which corresponds to a flow of about 6×10^6 electrons (the electron charge $q = 1.6\times10^{-19}$ As) per second. Be aware, however, that electrons do not flow at equal time intervals. In an APD with M=100 in the linear mode, the dark current is 100 pA (the same volume for thermal generation and the same purity are assumed as for the PIN photodiode), i.e. a flow of about 6×10^8 electrons per second. In the SPAD, the peak avalanche current can be in the mA range (see, for instance, figure 1.46), which is far higher than electronic circuit noise and can be detected 'easily.'

Thermally generated electrons or holes may trigger Geiger-mode events, as may charge carriers tunneling through narrow potential barriers (at highly doped p–n junctions). These two effects happen in darkness. An SPAD can therefore fire without photons. Thermally generated or tunneling charge carriers determine the dark count rate of an SPAD. For an SPAD low-field reverse current of 1 pA, 6×10^6 charge carriers are available per second. If we assume a PDP of 10%, that equates to a DCR of 6×10^5 dark counts per second.

When we assume self-quenching, i.e. the SPAD is floating (not connected to a quenching resistor or a quenching transistor), the capacitance of the SPAD, C_{SPAD}, is discharged by the avalanche current down to its breakdown voltage, where the

SPAD quenches itself, and it is better to consider avalanche charge. The avalanche charge Q_{av} is given by

$$Q_{av} = C_{SPAD}V_{EX}. \tag{1.26}$$

This avalanche charge flows through the SPAD. Some of these charge carriers can be trapped at impurities or defects inside the SPAD. For larger avalanche charges, more traps are filled. This is important, because these trapped charge carriers are released statistically afterwards and can trigger Geiger-mode events, i.e. the SPAD can fire with an uncertain delay after a 'true' Geiger-mode event, i.e. one that was caused by photon absorption, or by a dark count. These traps are said to be responsible for so-called afterpulses. The afterpulse probability is proportional to Q_{av}. Although wafers and silicon technologies for chip fabrication are quite pure nowadays, the DCR and APP are still present and limit the performance of SPAD sensors. The DCR and APP also increase with the area of an SPAD. Filled trap states decay exponentially with time and the APP can, therefore, be reduced by longer dead times, which means that the SPAD is not recharged immediately after a Geiger-mode avalanche but after a dead time (usually 10 ns or longer, since the decay time of filled traps is on the order of a few ns).

1.2 Discrete dedicated SPADs

This chapter describes SPADs designed for dedicated sensor technologies, which are often used in works described later. We use 'dedicated' in the sense that the fabrication technology is optimized in such a way that the physics of avalanche photodiodes is exploited to the greatest possible extent. This best exploitation of impact ionization is only possible for discrete SPADs. CMOS and BiCMOS processes do not allow for the necessary process modifications. The device physics determines that the impact ionization coefficient of electrons is larger than that of holes. The second important fact is the exponential decay of photogeneration with depth inside the silicon. In the next section, we will learn how these facts influence the structure of dedicated or customized SPADs.

1.2.1 Dedicated SPADs

In reference [7], a double-epitaxy SPAD was presented. The planar structure (see figure 1.4) consists of an n+/p+ junction situated in a p-type epitaxial layer. The cathode and anode contacts are at the surface. A p+ buried layer is implemented to provide a low ohmic path to the side contacts. An n-type substrate separates the device from the rest of the implemented structures.

The breakdown voltage of this device is 28.7 V and the maximum PDP of a 50 μm-diameter device is 48% when exposed to 550 nm light at an excess bias of 5 V. This work was commercialized by Micro Photon Devices S.r.l. (MPD) as a photon counting and timing module [8]. The p+ avalanche layer and the p-epitaxial layer are chosen to exploit the high electron impact ionization coefficient, since the electrons photogenerated in the p-epitaxial layer drift upwards into the multiplication zone and have the whole thickness of the multiplication zone available for

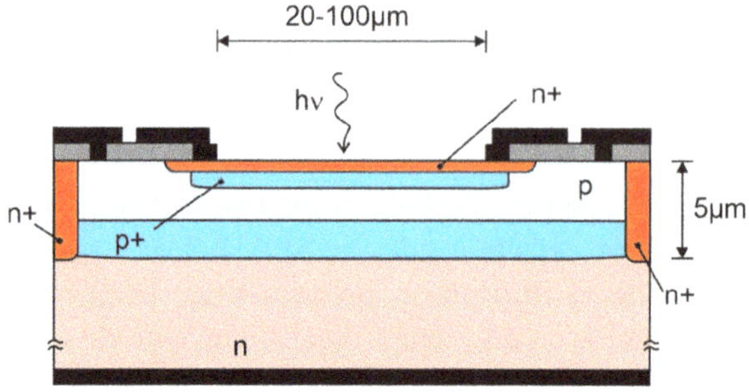

Figure 1.4. Cross section of a double-epitaxy SPAD (not to scale) [7].

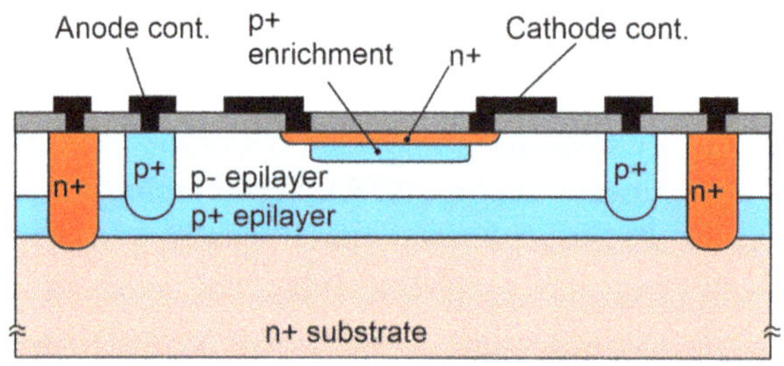

Figure 1.5. Cross section of a red-enhanced RE-SPAD [9].

impact ionization. This increases the PDP, but the limited epitaxial-layer thickness of 5 μm causes the PDP maximum to occur at 550 nm due to the increase of the $1/e$ penetration depth with the wavelength. The n-type substrate allows the suppression of electrical crosstalk between SPADs and individual adjustment of the breakdown voltage and excess bias for each SPAD, if arrays are used.

An improvement in the detection of red light was reported in [9]. The principal cross section of the improved, so-called red-enhanced SPAD (RE-SPAD) can be seen in figure 1.5. Again, the structure is a double-epitaxial device, but the epitaxial layer is thicker than before. We can retrieve an epitaxial-layer thickness of about 10 μm from figure 1.6.

In [9], not only was the thickness of the epitaxial layer increased, but it was also engineered to optimize the breakdown voltage. With unchanged doping profiles but an extended epitaxial-layer thickness, a breakdown voltage V_{BR} of close to 200 V was simulated, which is undesirable due to the fact that the power dissipated during an avalanche is rather high and the major part of it is dissipated in the drift zone. Therefore, the authors individualized the doping profiles, as depicted in figure 1.6 of the epitaxial layer, to reach a low electric field in the drift region and thereby reduced V_{BR} to approximately 60 V.

A back-illuminated SPAD is used by the Excelitas counting modules [10]. Figure 1.7 shows the cross section of this structure, which was already developed by the 1980s [11, 12]. On the front, the cathode is formed by an n+ area with a p enrichment zone that forms the electrical field. The n− guard ring avoids edge breakdown. Since the device is illuminated from the back, the contact of the cathode

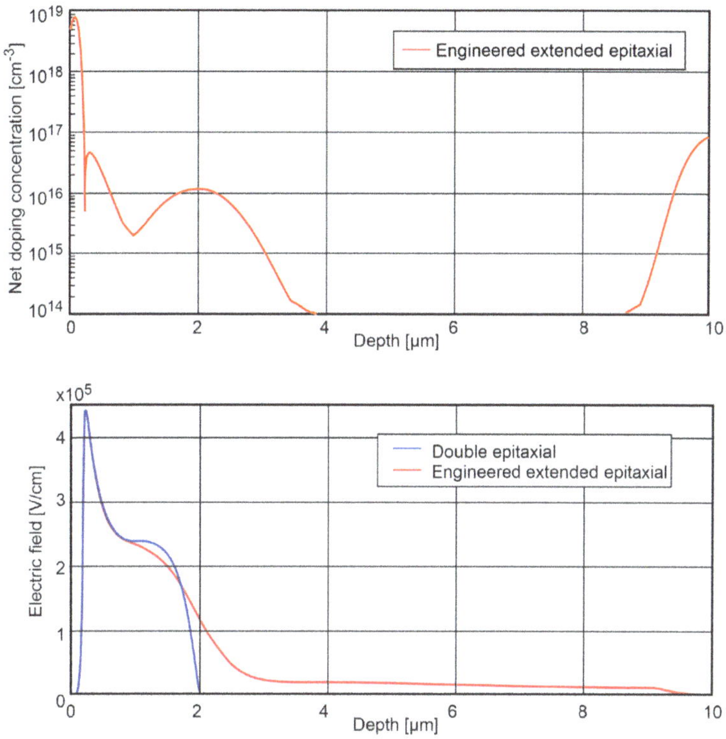

Figure 1.6. Doping profile and electric field of an RE-SPAD [9] 2012, reprinted by permission of the publisher (Taylor & Francis Ltd, http://www.tandfonline.com).

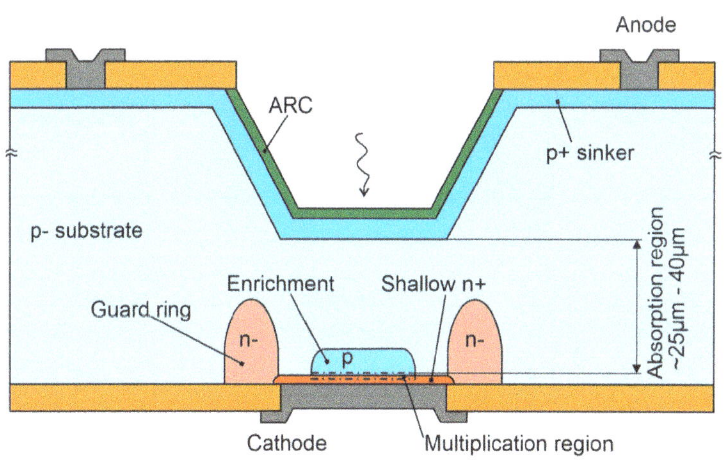

Figure 1.7. Cross section of a back-side illuminated SPAD by Excelitas Technology, formerly RCA Electro-Optics [14].

is directly on top of the n+ layer; its metallization can be used for back-reflection, thereby enhancing the detection efficiency [13]. After flipping, the device is etched from 40 μm down to 25 μm [11], and the p+ anode is formed in the p− quasi-intrinsic epitaxial substrate. An antireflective coating (ARC) enhances the detection efficiency further. This structure leads to a thick absorption region in the quasi-intrinsic epitaxial substrate and a multiplication region around the n+/p junction at the cathode. The electrons photogenerated in the absorption region drift downwards. The concentration of photogenerated electrons is large at the p+ surface and small inside the multiplication zone for the back-side illumination. As a consequence, more electrons have the full thickness of the multiplication zone available for impact ionization than would be the case for front illumination (from the bottom). The depletion of the thick absorption region leads to a breakdown voltage of about 400 V [12].

The high purity of the process used enables a very low dark count rate of a minimum of 25 cps (for SOCM-AQRH-W6 [10]), although the active volume is large. The PDP spectrum is depicted in figure 1.9.

Laser Components [15] also uses the back-side illuminated approach as described in [13, 16, 17]. The structure is similar to the one above and that depicted in figure 1.8. The low-doped π region is only 25 μm thick and therefore has a breakdown voltage of about 125 V at room temperature, which is rather low for a reach-through diode, for which it would normally be above 250 V. To obtain a longer photon conversion path, the top and part of the bottom surface are covered with metal to reflect photons. The PDP spectrum is shown in figure 1.9. The different types of Laser Components diodes start with a dark count rate of 10 cps [15] and have a very low afterpulsing probability of 3.2% at an excess bias of 15 V and a dead time of 24 ns at −10 °C [16].

Figure 1.9 shows a comparison of the PDPs for different types of custom technology SPAD. The RE-SPAD [9] shows the improvement of the PDP for red light, compared to the thinner MPD SPAD from [8]. The PDP enhancement of the EXCELITAS [10] and Laser Components SPADs [15] in the red and near-infrared spectral range is due to the thicker epitaxial layer, compared to those of the other

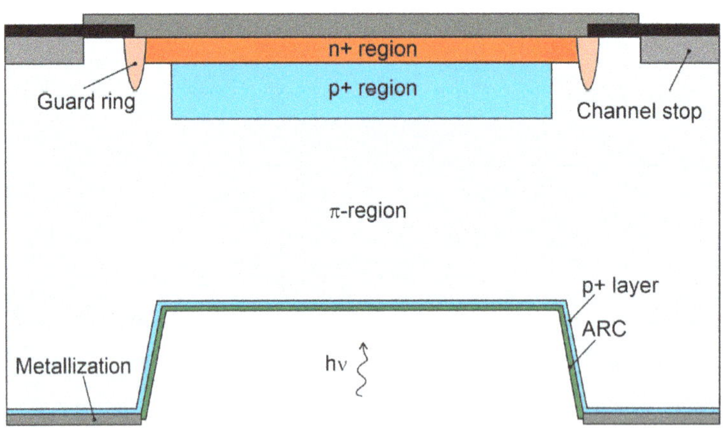

Figure 1.8. Cross section of back-side illuminated SAP500 by Laser Components [17].

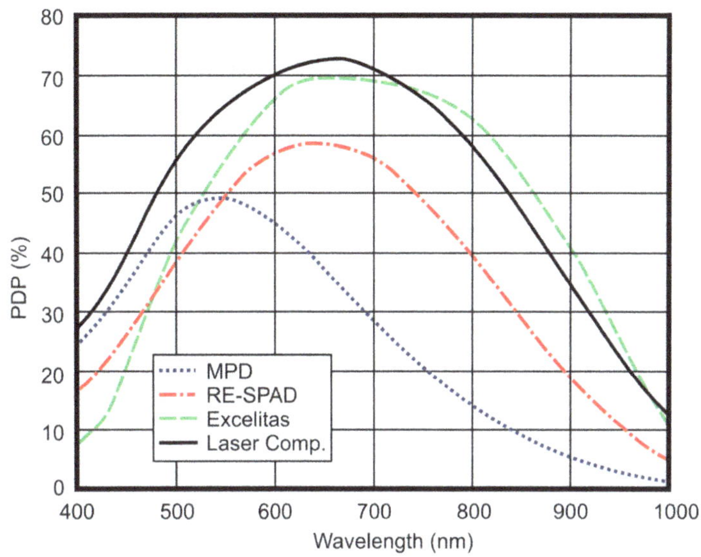

Figure 1.9. Comparison of the PDPs of different devices for different wavelengths; dotted: MPD [8], dash-dotted: RE-SPAD[9], dashed: Excelitas [10], solid: Laser Components [15].

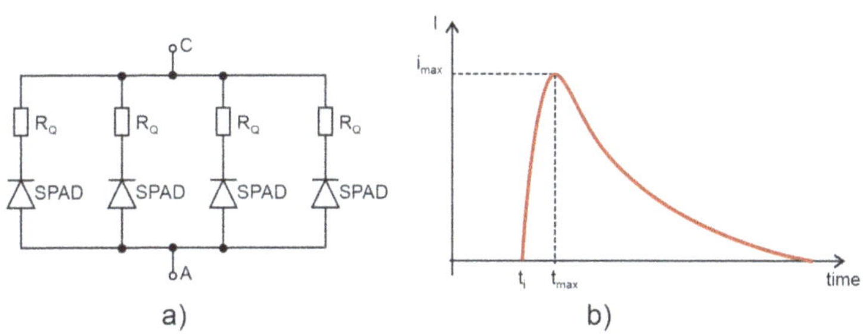

Figure 1.10. (a) Schematic of a silicon photomultiplier and (b) the output current of a single cell.

SPADs. This leads to a thicker absorption region, resulting in better detection of the longer wavelengths, which have larger penetration depths. All products work with an optimum excess bias voltage, which is not given in the data sheets and therefore not mentioned in the figure.

1.2.2 Silicon Photomultipliers

The so-called silicon photomultiplier (SiPM) is a device in which many SPADs with quenching resistors are connected in parallel; see figure 1.10(a). Each SPAD generates an output current pulse (see figure 1.10(b)) when hit by a photon and the output pulses are superimposed at the output nodes. Therefore, the number of detected photons can be estimated from the signal height. Nevertheless, this is only true for hits in different pixels; if one pixel is hit by more than one photon, the output does not vary. The behavior of SiPMs is similar to that of bulky photomultiplier tubes. However, SiPMs need much smaller reverse voltages and are insensitive to magnetic fields. Positron-electron tomography (PET) is therefore a large application field for SiPMs; it is also combined with magnetic resonance tomography (MRT). Most SiPMs are optimized for the blue light emission from scintillators used in PET. However, there are SiPMs for visible light and even near-infrared (NIR) light detection, e.g. the RGB-HD SiPM from Fondazione Bruno Kessler (FBK) [18] and their device with NIR-extended sensitivity [19].

Figure 1.11 shows a principal cross section of a SiPM as well as the corresponding top view [20]. It can be seen that the anodes of the single cells are built from the p substrate, while the cathodes and quenching resistors R_Q are connected together by metal contacts. This simple structure makes it possible to integrate several hundreds or even several thousands of single cells together, depending on the size of a single cell and the scope of the application.

The SiPM does not store any charge or generate outputs for each pixel like the array structures in single- or multi-channel devices. It produces an analog transient output in real time.

A simplified equivalent circuit is depicted in figure 1.12. C_J represents the junction capacitance of the SPADs, R_Q is the quenching resistor, and R_S represents the series resistance of the SPAD structure during discharge. Under ideal conditions, which means no dark counts, the switch S is open before a photon hits the device. The SPAD is prebiased to V_{bias}, which is larger than the breakdown voltage V_{BR}. When

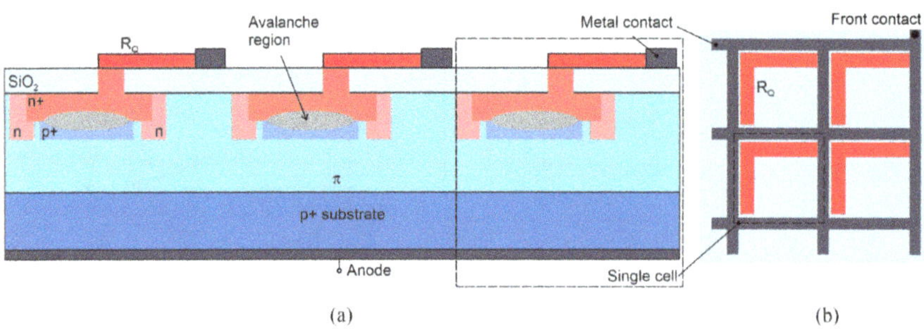

Figure 1.11. Principal structure of a SiPM: cross section on the left-hand side and top view on the right [20].

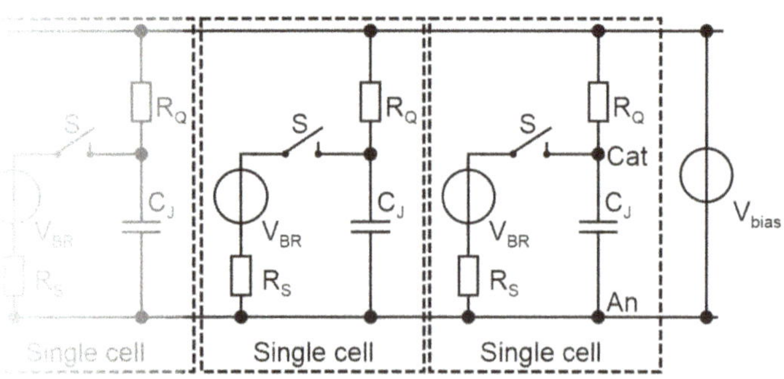

Figure 1.12. Simplified equivalent circuit for a SiPM [20].

a photon is detected, the switch S is closed, and the voltage across the SPAD drops to V_{BR} during the discharge.

$$i_{max} = \frac{(V_{Bias} - V_{BR})}{(R_Q + R_S)} \tag{1.27}$$

The output current i_{out} rises because the voltage at the cathode of the SPAD equals the bias voltage, which is proportional to $1 - e^{-\frac{t}{R_S C_J}}$, since $R_S \ll R_Q$. At t_{max}, the maximum current i_{max} is reached. From equation (1.27), it can be seen that the maximum current depends on the excess bias voltage $V_{EX} = V_{bias} - V_{BR}$. S opens and the SPAD is recharged by V_{bias}, the current falls proportionally to $e^{-\frac{t}{R_Q C_J}}$; see figure 1.13.

The gain of the SPAD can be calculated via the charge generated by one electron, given by equation (1.28).

$$\mu = \frac{Q}{e} = \frac{i_{max} \cdot \tau}{e} = \frac{1}{e}\frac{V_{EX}}{R_Q + R_S} \cdot R_Q C_J \tag{1.28}$$

If we assume again that the internal resistance R_S of the SPAD is small compared to that of the quenching resistor R_Q, then the gain does not depend on the quenching resistor:

$$\mu \cong \frac{V_{EX} \cdot C_J}{e} \qquad \text{for } R_S \ll R_Q \tag{1.29}$$

For more than one cell firing at the same time, the output currents superimpose to form a higher current.

A slightly more detailed SPAD model was published in [21], see figure 1.14.

The quenching resistor is represented by its resistance R_q and a parallel parasitic capacitcance C_q, which were determined to be 300 kΩ and 8 fF, respectively. The SPAD itself is represented by the resistance of the diode R_d, which was given as 1 kΩ and the junction capacitance C_d, which was given as 80 fF. To also include metal

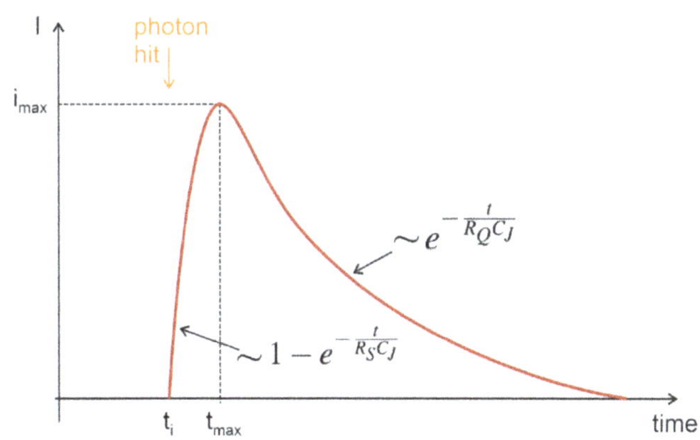

Figure 1.13. Output pulse of a SiPM [20].

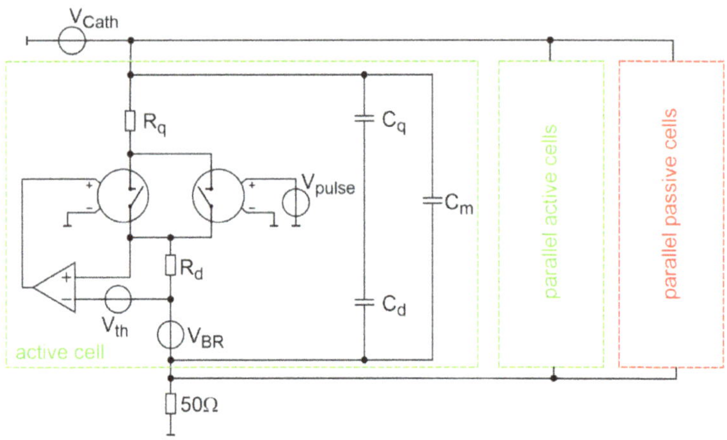

Figure 1.14. SiPM model after Marano [21].

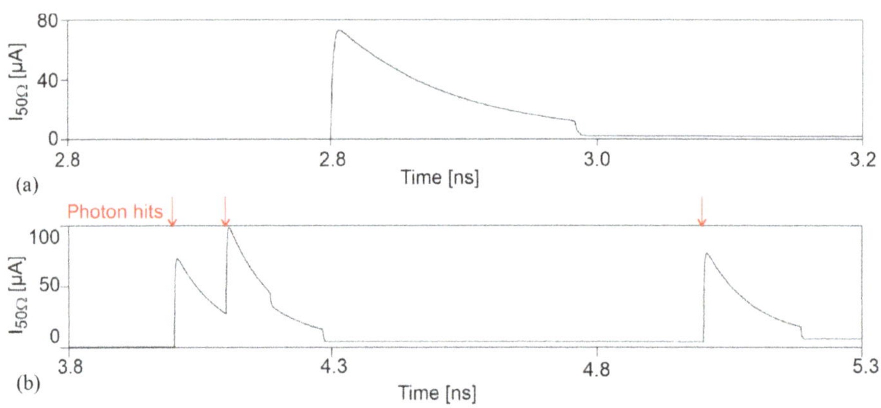

Figure 1.15. Currents of (a) single-photon detection and (b) multiple photon detection by a SiPM through a 50Ω resistor.

line capacitances and other parasitic capacitances, C_m was included at 0.5 fF. The values given in [21] were experimentally validated and considered for one firing element. As soon as there are several cells firing, the circuit can be modified to combine all triggered circuits together by dividing the resistors by the number of active elements and multiplying the capacitances accordingly. Furthermore, the switch representing an avalanche event was improved. It consists of two parts: one opens only for a short time for the initial photon hit and the other parallel switch holds the avalanche active until a threshold value of the current through the diode is reached, and the SPAD is quenched.

Figure 1.15(a) depicts a pulse from a single cell passing through a 50Ω load resistor. Again, it can be seen that the current increase is steep (see figure 1.13), and it is quenched when a certain value of the current is reached. There is a current tail of about 3.5% of the maximum current, which slowly discharges according to the time constant of the quenching resistor and the junction capacitance C_d.

Figure 1.15(b) shows the simulated current through the load resistor for the case of multiple hits at different times. The simulation was performed in LTspice®.

Several commercially available products are already on the market. For example, SiPMs manufactured by Hamamatsu Photonics [22] have been used in numerous studies. The S14160 series [23] offers fill factors of 31% and 49% for different types, respectively, as well as a large number of cells, up to nearly 90 000. The peak sensitivity is reached at 460 nm for all types of this series and the corresponding PDE is 18% for versions with the lower fill factor and 32% for those with the higher fill factor. All the photomultipliers have a breakdown voltage of about 38 V and the recommended excess bias voltages are 5 and 4 V, respectively. The DCR depends on the active area and varies from a maximum of 360 kcps for the device with the smallest active area to 2100 kcps for the largest. These devices are intended for surface mounting.

Improved mounting was realized in the S14160/S14161 series [24]; it offers tile-like mounting in all four directions and requires only a 0.1 mm gap between one active area and the next. The S14161 even offers up to 64 SPADs on each device. The recommended excess bias voltage is 2.7 V and the peak PDE is 50% at 450nm. The gain of the SiPM is given as 2.5×10^6.

The S13360 series is another product series [25]; it offers fill factors of up to 82%, a PDE of up to 50% at 450nm for an excess bias voltage of 3 V, and a maximum DCR of 6000 kcps for the 6400 pixel device. The crosstalk probability is 7%, and the reported gain is $4{\times}10^6$. The breakdown voltage for this series is higher than that noted above, at 53 V ± 5 V. All the described devices are sensitive to visible light. Additional cooling is offered by the 13362 series [26], which reduces the DCR and the APP.

These SiPMs are connected to a transimpedance amplifier in the so-called module series from Hamamatsu [27]. Figure 1.16 shows the basic circuit of these modules. The current pulses at the output of the SiPM are rather high due to the high gain of the SPAD devices; therefore, there is less need to increase the gain much more with the amplifier, which leads to more freedom on the circuit side.

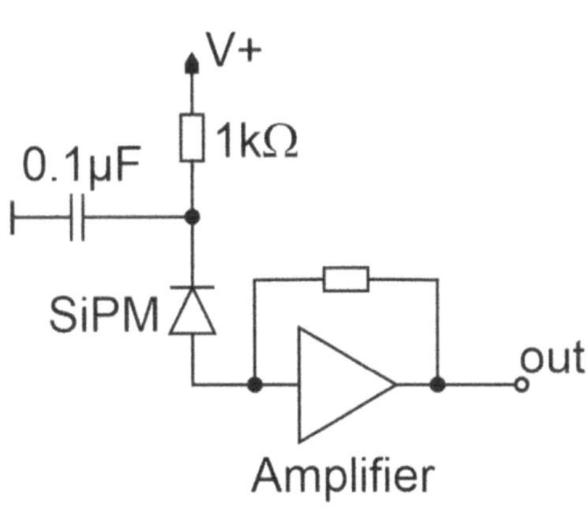

Figure 1.16. Basic circuit of a SiPM module. [27].

These modules are available with digital or analog outputs and cooled or uncooled. Depending on the module type, the −3 dB bandwidth of the modules lies between 2 MHz [28, 29] and 5MHz [30]; the optical behaviour depends on the SiPM used, as described above. The spectral sensitivity varies from visible (VIS) to NIR [27].

1.3 SPADs integrated into CMOS and BiCMOS

1.3.1 Thin SPADs

The following section describes what we call thin SPADs. Figure 1.17 depicts a principal example of a cross section of a thin SPAD. The absorption and multiplication zones are situated in a deep n-well. These devices typically have high PDPs in the blue to green wavelengths due to the penetration depth of the light and the rather shallow absorption zone. The device is separated from the substrate by a reverse-biased p–n junction between the deep n-well and the p-substrate, and can therefore be completely isolated. The anode is connected to the circuitry and the cathode is biased with a high positive voltage to apply breakdown and excess bias voltages. Carriers photogenerated in or below the space-charge region at the deep n-well/p-substrate junction are lost to the positive SPAD supply and are not multiplied by the avalanche effect.

SPADs with active diameters of up to 500 μm were reported in a 0.35 μm high-voltage CMOS technology [31]. The cross section and electric field of this SPAD can be seen in figure 1.18. The n-enrichment area constitutes the avalanche region. A maximum PDP of 55% was reported at 450nm for an excess bias voltage of 6 V. Interestingly, the temperature characterization of the breakdown voltage, which varies between 23.5 V at −50 °C and 26.5V at + 50 °C, changes by less than ± 6% compared to its value at room temperature. The APP was also measured for different SPAD diameters by determining the hold-off time required to reach a 1% APP. It rose from 40 ns for a 20 μm diameter to 100 ns for 100 μm and even 150 ns for a 500 μm diameter.

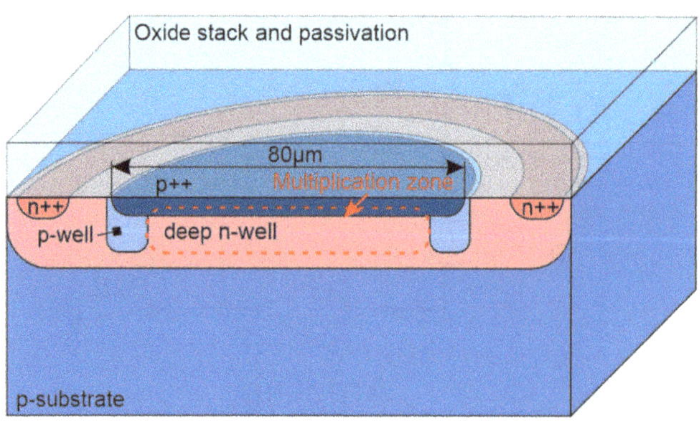

Figure 1.17. Principal cross section of a thin SPAD.

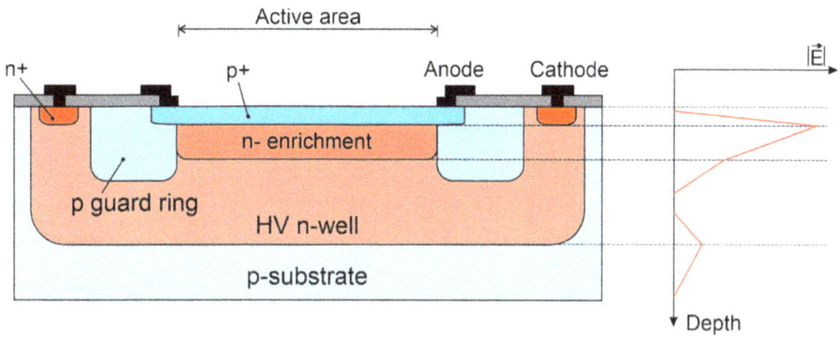

Figure 1.18. Cross section and electric field of SPADs with diameters of up to 500 μm [31].

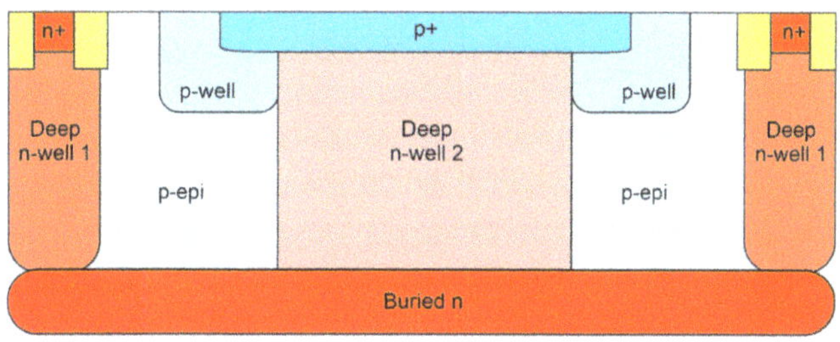

Figure 1.19. Cross section of the 180 nm CMOS SPAD [32].

Reference [32] presented an SPAD in a 180 nm CMOS technology. The cross section of the circular 12 μm-diameter device is depicted in figure 1.19. The device was isolated from the substrate by a buried n layer. A p-epitaxial layer formed a guard ring to avoid edge breakdown and to increase the possible excess bias, since the guard ring breakdown voltage was 12.2 V higher than the breakdown voltage of the main junction, which was 23.5 V. With $V_{EX} = 10$ V, a DCR of 12.84 cps per μm^2 was reported and the corresponding PDP was 47.6% for 480 nm light; between 440 nm and 620 nm, the PDP was greater than 40%. An APP of less than 0.3% was reported for a dead time of 300 ns.

Reference [33] presented an integrated SPAD combined with passive quenching and an active reset circuit in a 0.18 μm CMOS image sensor technology. The image sensor technology allowed higher doping levels at greater depths, increasing the PDE of carriers generated deeper in the well; in particular, the layer labeled N-SPAD shows more doping at a greater depth than would be the case for n+ in standard CMOS technologies. The cross section of the proposed SPAD is depicted in the second part of the figure (figure 1.20). The diode (which has no quenching circuit) shows a PDE of 16.45% for a fill factor of 35%, corresponding to a PDP of 47% at 580 nm at an excess bias voltage of 5 V. The DCR for the same V_{EX} is 380 Hz and the breakdown voltage of the device is about 20 V.

The advantages of 180 nm CMOS image sensor technology with retrograde deep n-wells were presented in [34]. These retrograde wells have a low dopant level at the surface, which increases to a maximum deeper in the silicon. The low dopant level

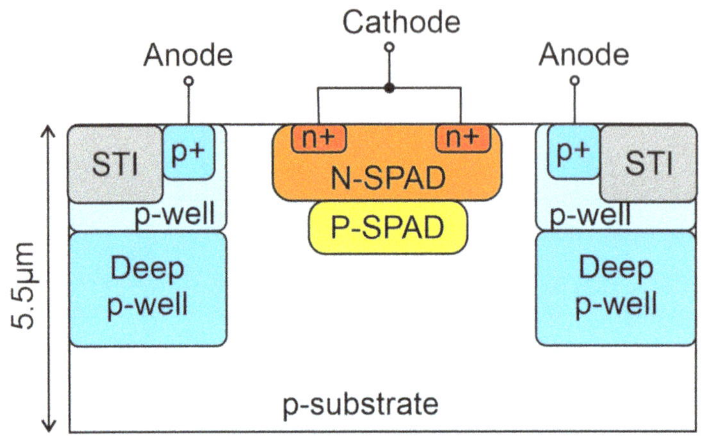

Figure 1.20. Cross section of the integrated SPAD [33].

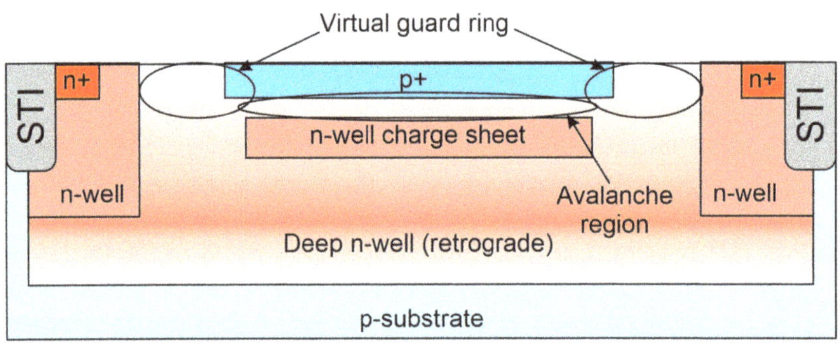

Figure 1.21. Cross section of an SPAD in 180 nm CMOS image sensor technology [34].

close to the surface allows the formation of a virtual guard ring around the active area and avoids edge breakdown. The active area is circular and has a diameter of 12 μm. A p+ region is located at the top; below it is an n-type charge sheet, which is formed by ion implantation; see figure 1.21.

The multiplication region is located between the p+ region and the n-type charge sheet. The doping levels allow the electric field to persist into the deep n-well layer; therefore, charge carriers are collected over a wide depth range, which enhances the PDE. The PDEs are around 15% and 20% at a wavelength of 400 nm for excess bias voltages of 1 V and 3.3 V, respectively, and increase to plateaus at about 34% and 46%, respectively, between 425 nm and about 500 nm. For longer wavelengths, the PDEs decrease to 19% and 17%, respectively, for 700 nm and finally drop below 15% for both excess bias voltages for 800 nm. The maximum V_{EX} is obtained at 3 V, for which the breakdown voltage is 21.4 V. The DCR is 140 cps at room temperature.

Another work that used 180 nm CMOS technology was presented in [35]. The cross section of the SPAD can be seen in figure 1.22; these circular diodes were

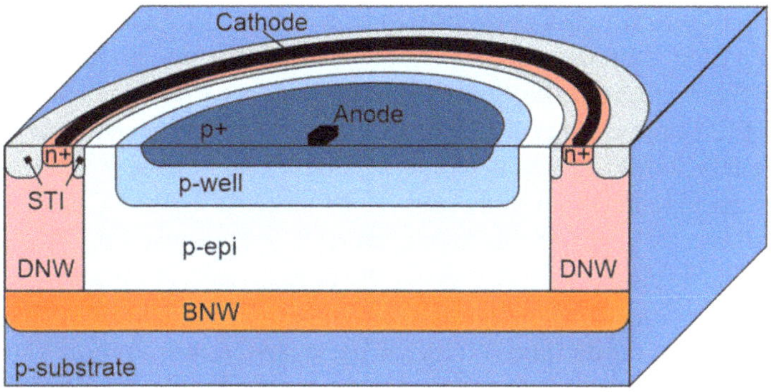

Figure 1.22. Cross section of the 180 nm CMOS SPADs with diameters from 25 to 100 μm [35].

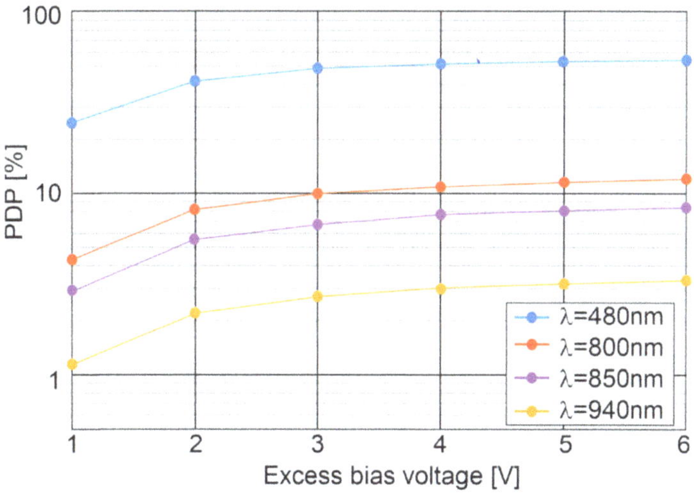

Figure 1.23. PDP for different wavelengths versus excess bias voltage. Copyright 2009 IEEE. Reprinted with permission, from [35].

produced in three different sizes from 25 to 100 μm in diameter. A p-well anode and a buried n-layer cathode allow for a p-epitaxial layer in between them, which generates a deep high-field region and therefore improves the sensitivity spectrum. The maximum excess bias was 6 V and a breakdown voltage of 22 V was reported at room temperature. The median DCR was 0.2 cps per μm^2 for $V_{EX} = 6$ V. In this work, the temperature behaviour was also investigated. It was shown that the DCR decreases to 1.6 mcps per μm^2 at −65 °C for the 25 μm device. The smallest device was used for the measurement of the APP; with a V_{EX} of 6 V and a dead time of 11 ns, an APP of 0.1% was measured. The maximum PDP of 55% was obtained for 480 nm light, saturation of the PDP was reported to start at an excess bias voltage of 5 V, and therefore an insensitivity to breakdown voltage variation was mentioned [35]. Figure 1.23 shows the PDP dependence on the excess bias voltage for different wavelengths.

Two types of thin SPAD were reported in a 180 nm HV CMOS technology [36]. The authors stated that they concentrated on the 180 nm node as a compromise between, on the one hand, the demand for speed from the electronics and, on the other hand, the doping concentration and oxide stack, which the diodes want to avoid and which would be even worse in smaller feature-size technologies. Both devices were circular and had diameters of 20 μm; see figure 1.24.

The breakdown voltages of the devices were 49.9 V and 82.1 V for types A and B, respectively. The doping concentration of the deep p-well (DPW) layer is higher than that of the high-voltage p-well (HV PW), which leads to the reduced breakdown voltage of the A-type diode. The DCRs for these devices were reported to be 0.68 cps per μm^2 and 1.06 cps per μm^2 for A and B, respectively. For the PDP, the different breakdown voltages were taken into account. For better comparability, the excess bias voltage was related to the breakdown voltage. The maximum PDP of both devices was obtained for 570 nm light. The peak PDE for 15% of V_{EX} was 22% for the A-type device, and for the B-type it was 19%. The saturation of the increase of the PDP was more pronounced for the B-type device.

Reference [37] presented an SPAD in a 130 nm CMOS imaging process that uses shallow trench isolation (STI) combined with a p-type passivation implant surrounding the STI to realize an effective guard ring in order to avoid edge breakdown. Since the doping concentration is high close to the STI surface, the presence of a free path for minority carriers is prevented, and therefore, the probability that carriers enter the active area is reduced. The doping concentration of the passivation implant decreases with increasing distance to the STI; this avoids

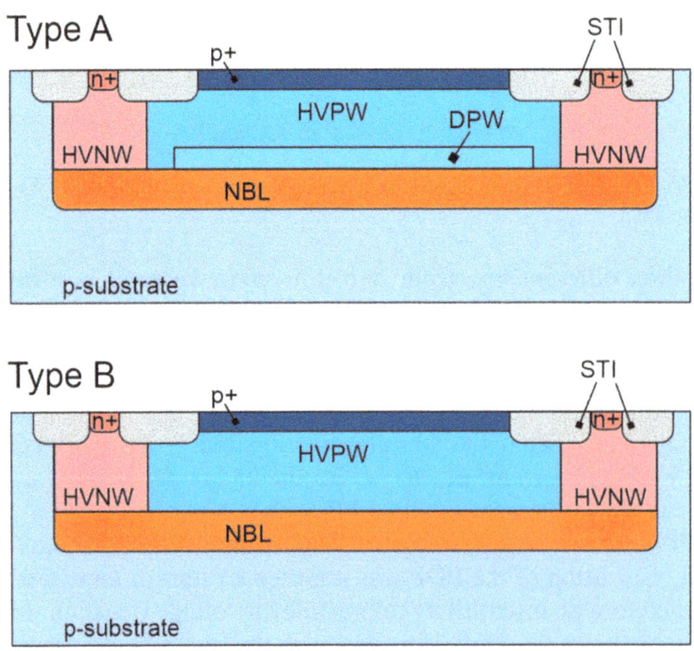

Figure 1.24. Cross sections of two types of SPAD: type A (top) and type B (bottom) [36].

electric field peaks and therefore edge breakdown. The cross section of this device is shown in figure 1.25. Since the doping concentration is high in sub-μm CMOS processes, the main source of dark counts is considered to be the tunneling effect. To minimize these tunneling events, the doping level of the n-well cathode was reduced in this work and the effects were compared. The device with standard doping showed a breakdown voltage of 9.4 V, while the lightly doped device had a breakdown voltage 12.8 V for 2.5 times less n-well doping. Figure 1.26 depicts plots of the DCRs of both devices at room temperature. It can be seen that the DCR is reduced drastically, e.g. from 90 kcps to 10 cps for a V_{EX} equal to 1 V.

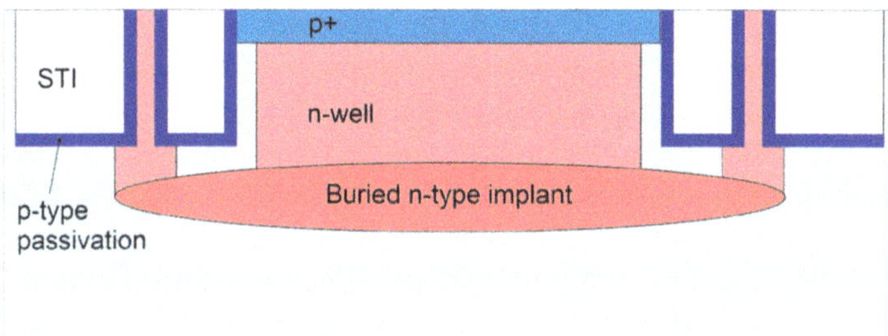

Figure 1.25. Cross section of a 130 nm imaging technology SPAD that uses the STI/passivation guard ring concept [37].

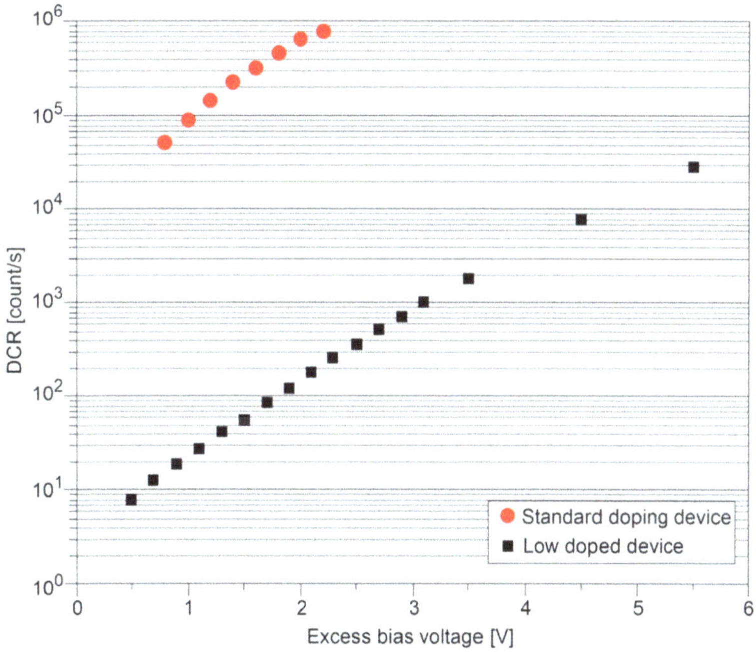

Figure 1.26. DCR versus excess bias voltage for differently doped devices in the 130 nm CMOS image process. Reprinted from [37]. Copyright (2009), with permission from Elsevier.

The PDPs were similar for both devices, which used the same optical stack. Due to the higher breakdown voltage of the lightly doped device, higher excess bias voltages were possible and therefore the PDP could be improved. The standard device reached a peak PDP of 28% at 480 nm with a V_{EX} of 2 V, while the lightly doped device reached 36% for a V_{EX} of 5 V. The corresponding DCR was 11 cps. The APP was zero for a dead time of 180 ns, which was reached by reducing the capacitance at the border of the detector. In addition, on-chip integration of the quenching circuit ensured minimum capacitance.

An SPAD with a diameter of 50 μm^2 that used the 130 nm imaging process was presented in [38]. The wells used for the device were the same as for the standard CMOS process. The imaging layers in the oxide stack above the diode improved the optical transmission. Figure 1.27 shows the cross section of the proposed SPAD. The anode was formed by a p+/p-well structure, while the cathode was formed by a deep retrograde n-well. This retrograde characteristic allows a low-dopant area close to the surface due to an implant stop that avoids the automatically generated wells. Due to the high dopant concentration and the shallow depth of the implanted wells as well as the shallow trench isolation, the DCR and APP were expected to be high. The breakdown voltage was 14.4 V and the maximum excess bias voltage was 1.4 V. The DCRs were measured over a wide temperature range for three different excess bias voltages: 0.6 V, 1.0 V, and 1.4 V. For the two lower bias points, the DCRs started to increase at around −4 °C and 0 °C, respectively. The DCRs increased rapidly with increasing temperature; measurements were reported up to 45 °C, at which the DCRs were above 100 cps for all three excess bias voltages. Below 15 °C, they were less than 20 cps. The APP at the same temperature was very low at a V_{EX} of 1 V due to a dead time of 100 ns and the minimization of the charge flowing during the breakdown. The peak PDE of 28% was reached for 500 nm light.

By adding microlenses on top of SPADs integrated in the 40 nm STMicroelectronics (STM) process node [39], the fill factor and the PDP of the devices were improved [40]. With an excess bias of 1 V and a breakdown voltage of about 14.6 V, only the bias voltage of the diode is given at 15.5 V, but the sensing threshold of the following inverting stage would have to be subtracted to calculate the breakdown voltage. However, this threshold was not specified in the paper [40]. The microlenses increased the PDP by about 10% in at the range of maximum sensitivity, from 450nm to about 550 nm, and additionally smoothed the curve.

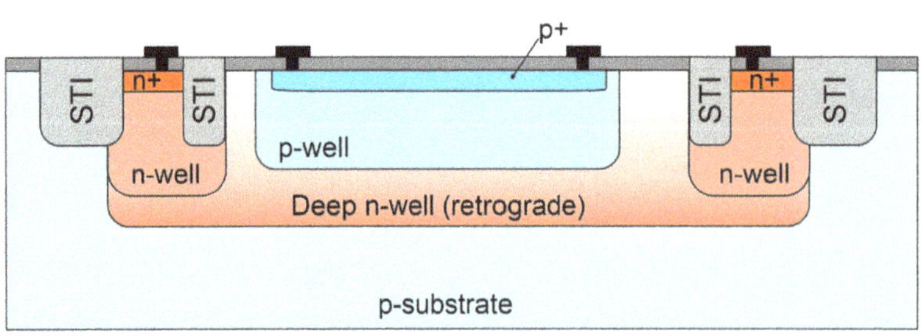

Figure 1.27. Cross section of the 130 nm CMOS imaging technology SPAD described in [38].

The maximum PDP was around 45% at 520 nm at room temperature with microlenses. It was reported that the PDP varied across the wafer by about ± 15%.

1.3.2 Thick SPADs

Figure 1.28 compares a thick SPAD to a thin SPAD. Thin SPADs have a thin combined multiplication and absorption zone within a deep n-well. The deep n-well is connected to a positive bias voltage that operates the thin SPAD above its breakdown voltage; carriers photogenerated in the p-substrate cannot trigger an avalanche. The thick SPAD has a thick lightly doped p-epitaxial layer below the p-well. The multiplication zone is located at the n+/p-well junction and reaches into the p-well. When the doping levels of the p-well and p-epitaxial layer are chosen appropriately, the p-well and the p-epitaxial layer can be completely depleted and carriers (electrons) generated in the thick epitaxial layer rapidly drift up to the multiplication zone and can trigger an avalanche. In addition, in silicon, the electron impact ionization coefficient is larger than the hole impact ionization coefficient. Therefore, a larger PDP is to be expected for a thick SPAD in the red to near-infrared spectrum than for a thin SPAD.

1.3.2.1 Thick SPAD in PIN photodiode CMOS

PIN photodiode (Bi)CMOS technology [42] can be used to implement an SPAD with a thick absorption zone when a p-type region, e.g. a p-well, is added below the n+ surface cathode to obtain a multiplication zone. Figure 1.29 shows the resulting thick SPAD in 0.35 μm PIN photodiode CMOS technology, which was introduced as a linear-mode APD in [43]. A structure was also investigated that had an n-well around the p-well (see the left part of figure 1.28) as a variation, in order to avoid edge breakdown [41]. The p-epitaxial layer had a thickness of about 12 μm. The breakdown voltage was about 28.7 V at 25 °C [43]. The p-well and the p-epitaxial layer were already completely depleted at 20 V, creating a thick absorption zone with a high carrier drift velocity.

The PDPs of several efficient SPADs are compared to the thick 0.35 μm PIN photodiode CMOS SPAD in figure 1.30. The thick SPAD in 0.35 μm PIN photodiode CMOS shows a similar or even better PDP for wavelengths longer than about 800 nm at an excess bias voltage of 6.6 V, instead of the 12 V excess bias described in [44]. Unfortunately, no antireflective coating can be implemented with the thick 0.35 μm PIN photodiode CMOS SPAD, and optical interference causes ripples in the spectral PDP curve for the 0.35 μm PIN photodiode CMOS SPAD in figure 1.30.

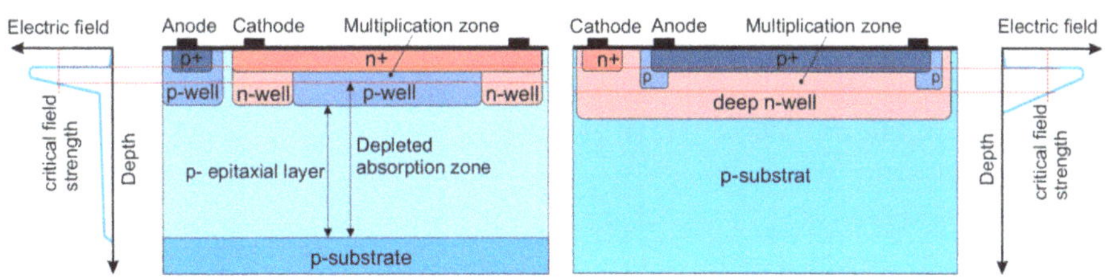

Figure 1.28. Comparison of thick and thin SPADs ([41] suppl. information. (2017) Copyright. With permission of Springer).

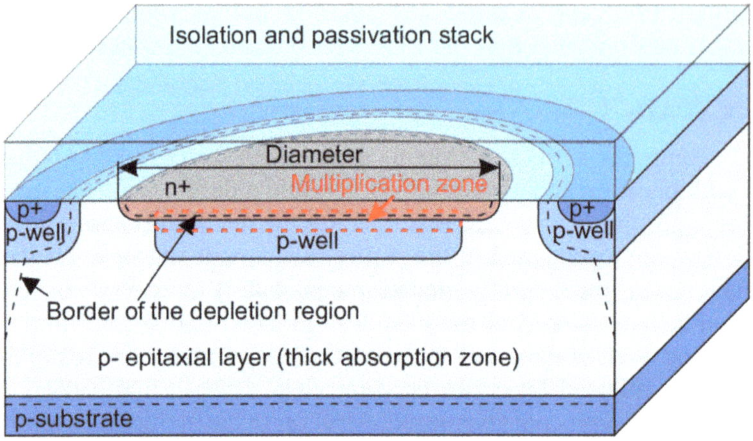

Figure 1.29. Structure of a thick SPAD in PIN photodiode CMOS technology.

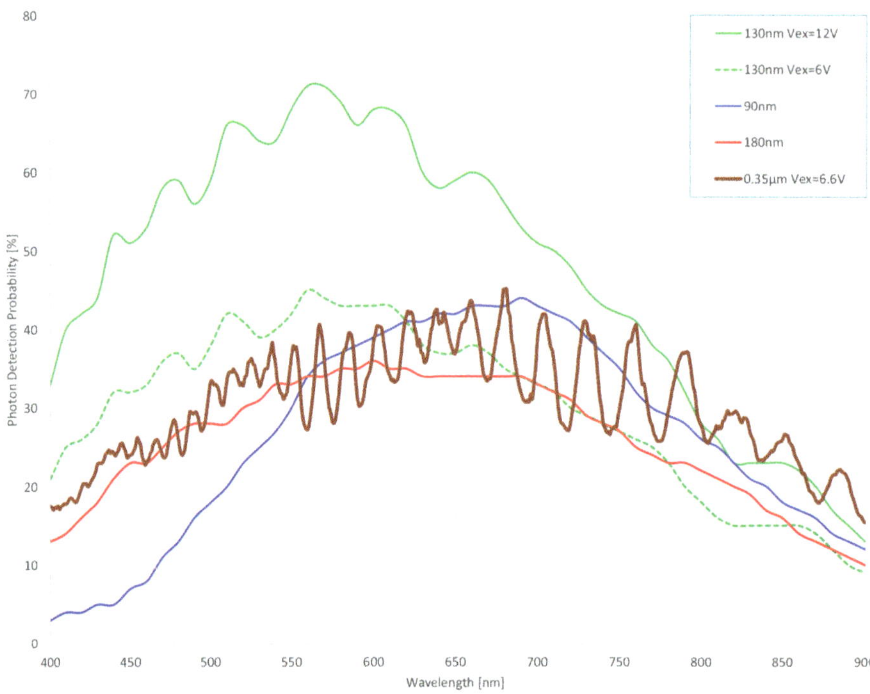

Figure 1.30. Comparison of the spectral photon detection probability of a 0.13 μm CMOS SPAD [44], a 90nm CMOS SPAD [45], a 0.18 μm CMOS SPAD [46], and the thick PIN photodiode CMOS SPAD in 0.35 μm technology.

Another thick SPAD was introduced [47, 48] with PDPs of 62.2% to 64.8% at 610 nm and an excess bias voltage of 5 V. The customized n-SPAD and p-SPAD regions were, however, not described.

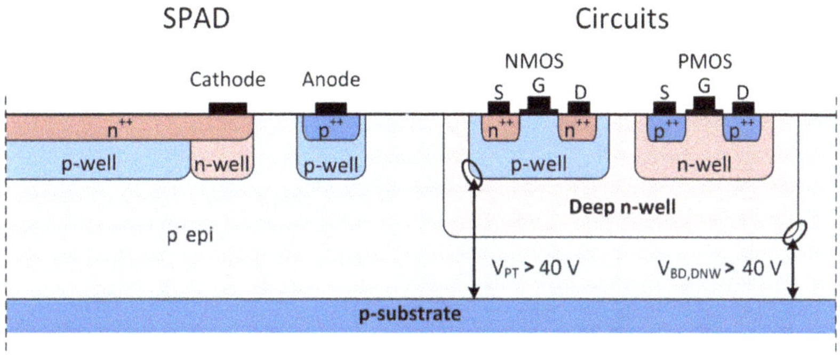

Figure 1.31. Isolation of thick SPAD and transistors on the same chip ([41] suppl. information. (2017) Copyright. With permission of Springer).

We now explain how a thick SPAD and circuits can be implemented on the same chip. Figure 1.31 shows the cross section of a thick SPAD in PIN photodiode CMOS together with an n-channel metal–oxide–semiconductor field-effect transistor (MOSFET) and a p-channel MOSFET. The transistors are surrounded by a deep n-well to isolate them from the negative substrate potential that is necessary for the high reverse voltage of the thick SPAD. The punch-through voltage V_{PT} is larger than 40 V from the p-well to the p-substrate for the CMOS process used, which means that for a p-well potential of 0 V, the p-substrate can be at −40 V, requiring a reverse SPAD voltage of a bit more than 40 V (e.g. 40 V + $V_{DD}/2$ if the cathode is connected to an input at $V_{DD}/2$). The deep n-well of the process used increases the breakdown voltage of the n-well toward the substrate. The breakdown voltage of the deep n-well toward the substrate, $V_{BD,\ DNW}$, is larger than 40 V. These breakdown voltages are a good fit for the breakdown voltage of the PIN photodiode CMOS SPAD.

Not only is the large PDP of SPADs interesting, but the parasitic properties, dark counts, and afterpulsing are also important for applications. These all depend on temperature. The DCR increases with temperature due to the thermal generation of electron–hole pairs, and the afterpulsing probability usually decreases with temperature (for a constant dead time), because charge carriers are released from traps earlier at higher temperatures. Both the DCR and the APP also depend on impurities (traps) and the interface states [49, 50]. The PIN CMOS SPAD, therefore, was investigated in the temperature range from −40 °C to 50 °C [51]. The cascoded active quenching circuit described in figure 3.4, which has a dead time of 9.5 ns, was integrated with an SPAD with an active diameter of 30 μm. A Thermonics T-2650BV allowed the temperature of the SPAD to be changed in the dark. Dark counts and afterpulses were separated using the interarrival-time histogram method (see figure 1.32).

Pulses that occur within 100 ns after a pulse were counted as afterpulses, from which the dark counts within this interval had to be subtracted. Pulses between 1 μs and 10 μs were used to determine the DCR value to be subtracted (the different periods were considered, of course). The dependence of the breakdown voltage on temperature was considered to keep the excess bias voltage constant for the different temperatures.

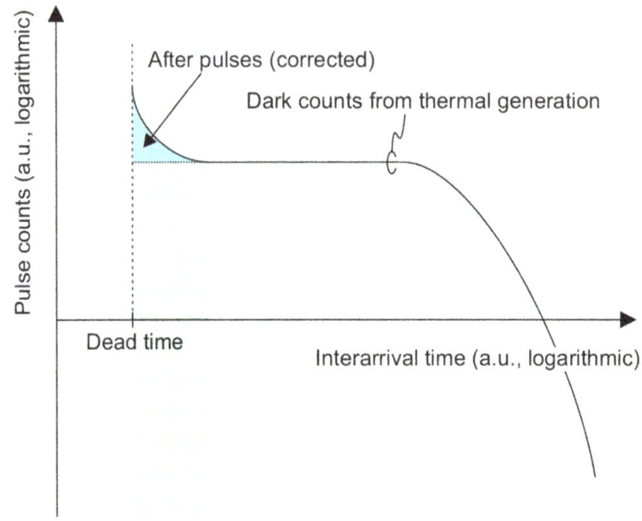

Figure 1.32. Interarrival-time histogram that uses the principle of determining the afterpulsing probability.

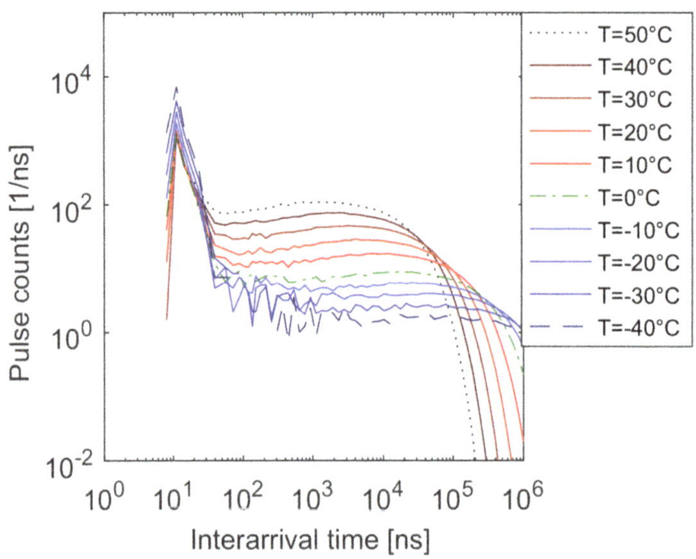

Figure 1.33. Obtained interarrival-time histograms at an excess bias voltage of 3.3 V for different temperatures. Reproduced from [51] with permission from Hindawi.

The results obtained for the arrival times are presented in figure 1.33. It can be seen that the DCR strongly depends on temperature.

The results for the dependence of the dark count rate on the excess bias voltage are shown for different temperatures in figure 1.34. When the DCRs for $V_{ex} = 3.3$ V are evaluated, the DCR increases by a factor of about 1.7 per 10 °C. In contrast to [50], in which the DCR did not reduce for temperatures lower than room

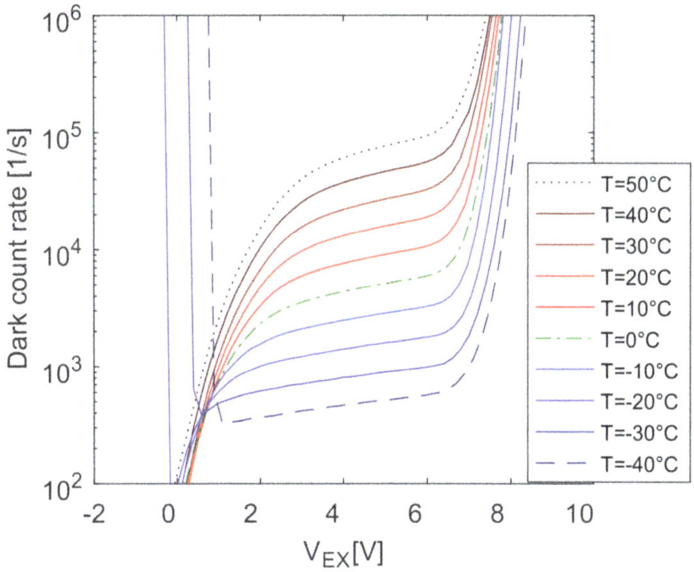

Figure 1.34. Dependence of the dark count rate on the excess bias voltage within the temperature range −40 °C to 50 °C. An excess bias of 0 V corresponds to the breakdown voltage, i.e. the (temperature dependent) breakdown voltage was subtracted from the total reverse voltage of the SPAD. Reproduced from [51] with permission from Hindawi.

temperature, this factor is almost constant from −40 °C to 50 °C. The very high DCRs for −40 °C and − 30 °C at excess bias voltages of less than about 1 V originate from oscillations of the active quenching circuit due to incomplete depletion of the epitaxial layer and the resulting capacitance increase of the SPAD [51]. Less than 1,000 dark counts per second are achieved for −30 °C and below for excess biases of up to about 6.6 V.

The results for the dependence of the afterpulsing probability on the excess bias voltage are shown for different temperatures in figure 1.35. The APP does not show a strong temperature dependence—which is unexpected with respect to the literature, if traps are the main cause of afterpulses. According to the literature, the release time for trapped charge carriers depends strongly on temperature: carriers are released much earlier for higher temperatures and much later for lower temperatures; therefore, the APP should increase when the temperature is decreased [50]. This is almost invisible in the APP results. In conclusion, trapped charges are not the main cause of afterpulsing in the investigated PIN CMOS SPAD. A possible explanation could be the emission of photons during an avalanche event, their absorption in the substrate, the diffusion of photogenerated electrons from the substrate into the depleted absorption layer of the SPAD after the dead time of the quencher, or the triggering of a new avalanche [51]. In contrast to the DCR, the APP differs much less within the temperature range from −40 °C to 50 °C.

A 0.35 μm PIN photodiode CMOS SPAD with a 50 μm diameter was investigated in [52] for further characterization. It was wire-bonded to the gating circuit [53] depicted in figure 4.20. Of course, the SPAD and the gater can be integrated together

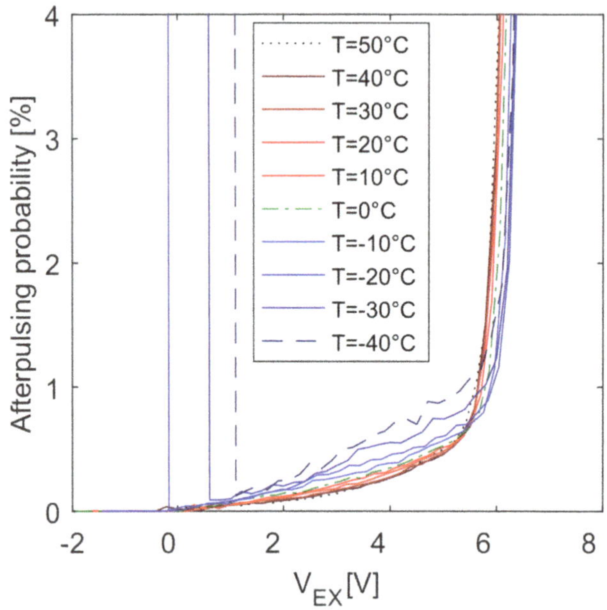

Figure 1.35. Dependence of the afterpulsing probability on the excess bias voltage within the temperature range −40 °C to 50 °C. An excess bias of 0 V corresponds to the breakdown voltage, i.e. the (temperature dependent) breakdown voltage was subtracted from the total reverse voltage of the SPAD. Reproduced from [51] with permission from Hindawi.

on one chip, but in these fabricated optoelectronic integrated circuits (OEICs), no pad was provided for a direct measurement of the dependence of the cathode potential on time. The only pads available were in a gater with an input pad fabricated for connection to discrete SPADs and a PIN SPAD with pads for characterization. The gater and the pads for wire bonding allowed the observation of the development of the avalanche after photon absorption. The gater charged the SPAD and allowed it to run freely. When a photon was absorbed, the avalanche started to build up, and because the cathode of the SPAD was floating, the avalanche current discharged the capacitance of the SPAD (and the capacitances of the bond pads between the SPAD and the gater). If no photon was absorbed during the gate window, the gater switched the SPAD off by reducing its reverse bias to the breakdown voltage or below until the next gate window started.

A 26 GHz picoprobe with an input capacitance of 50 fF, an 1.25 MΩ input resistance and an attenuation of 1:10 was placed on one of the two bond pads between the SPAD and the gater. The bond wire was short and the influence of its inductance on the transients was negligible. A halogen lamp was used as the light source.

Figure 1.36 shows the obtained transients of the cathode potential measured with the picoprobe and a 20 GHz/80 GS s^{-1} real-time oscilloscope. The clock frequency of the gater was 15 MHz. When the light was partly turned on (see the left part of figure 1.36), photon detection was distributed over the active gate time. When the light was fully turned on (see the right part of figure 1.36), the probability was high

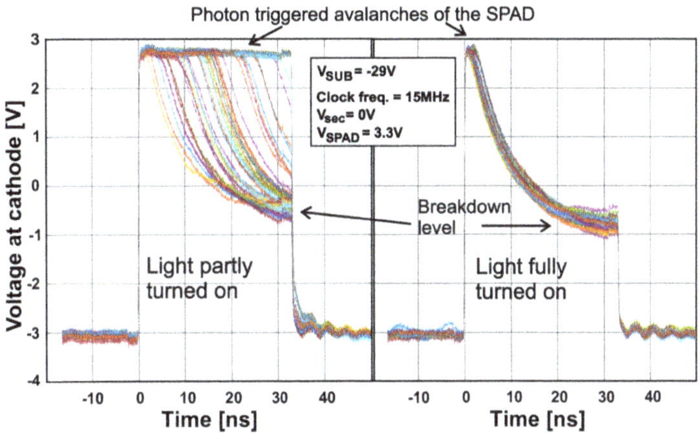

Figure 1.36. Voltage at the cathode of the PIN SPAD during self-quenching. ©2018 IEEE. Reprinted, with permission, from [52].

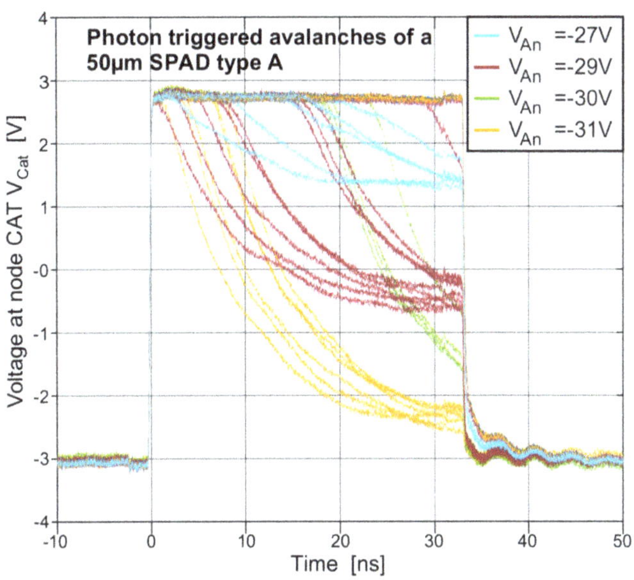

Figure 1.37. Cathode voltage transients of a PIN CMOS SPAD with a diameter of 50 μm for different anode voltages V_{AN}. In the illustration, the transients of several active phases are overlaid. It can be seen that early avalanches quench themselves (self-quenching) during the active phase when the breakdown level is reached. Some avalanches, which occur later in the active phase, are quenched by the gater in the following reset phase. The breakdown levels observed depend on the anode voltage V_{AN}. Reproduced from [54] with permission from MDPI.

that a photon was detected at the beginning of the active gate time (and the SPAD could not detect further photons, because self-quenching had happened already).

Figure 1.37 illustrates self-quenching down to the breakdown-voltage level for different anode voltages (substrate voltages). The gater clock frequency of 15 MHz corresponds to an active phase of about 33.3 ns and a reset phase of 33.3 ns (a duty

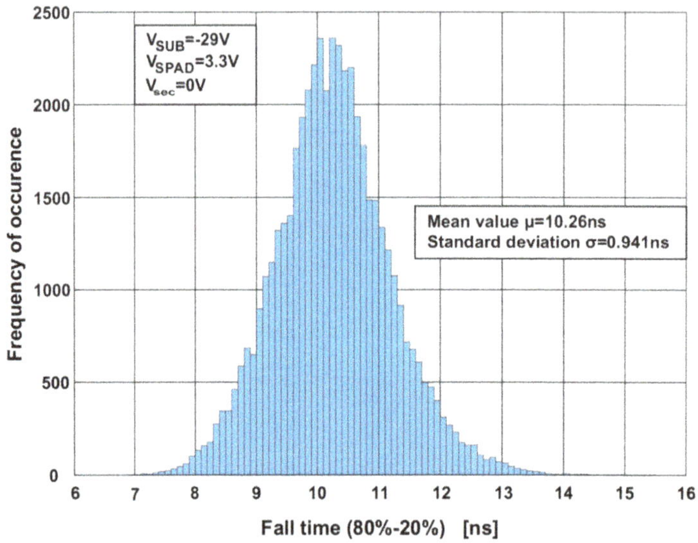

Figure 1.38. Histogram of the fall times of the PIN SPAD during self-quenching. ©2018 IEEE. Reprinted, with permission, from [52].

cycle of 50%). When the photon is absorbed early in the active phase, the self-quenching discharges the SPAD to the breakdown level V_{BR} due to the length of the 33.3 ns active phase; the breakdown voltage can be obtained from the voltage level at the end of the active phase V_{BD} by subtracting the (negative) anode voltage V_{AN} (e.g. $V_{BD} = -2.8$ V from the lowest yellow curve plus $V_{AN} = -31$ V gives a breakdown voltage of V_{BR}=28.2 V; the excess bias voltage is 2.9 V + 2.8 V = 5.7 V, where 2.9 V is the cathode voltage before the avalanche starts).

Figure 1.38 shows the distribution of the transient fall times for the self-discharge of the SPAD from 80% to 20% (the light is fully turned on and the excess bias is 3.6 V). The mean value of the fall times was 10.26 ns with a standard deviation of 0.941 ns [52]. It can be concluded that there is no strong dependence of the fall times (i.e. the avalanche build-up times) on the wavelength of the detected photons (the halogen lamp emits a wide spectrum).

The mean values and standard deviations of the fall times for different anode voltages (leading to different excess bias voltages) are shown in figure 1.39. 'Light off' corresponds to dark counts. There is no pronounced dependence of the fall time on the anode voltage or on the excess bias voltage. It could be the case that the two bond pad capacitances and the input capacitance of the picoprobe had a stronger influence on the avalanche build up than the excess bias voltage of the SPAD.

1.3.2.2 Thick SPAD in HV CMOS

A thick absorption zone was also realized in an HV 0.35 μm CMOS technology using epitaxial wafers [55]. Figure 1.40 shows the 3D structure of this SPAD. The standard epitaxial-layer doping of the order of 10^{15}cm^{-3} was partially compensated for by a deep n-well, which also partially compensated for the deep p-well.

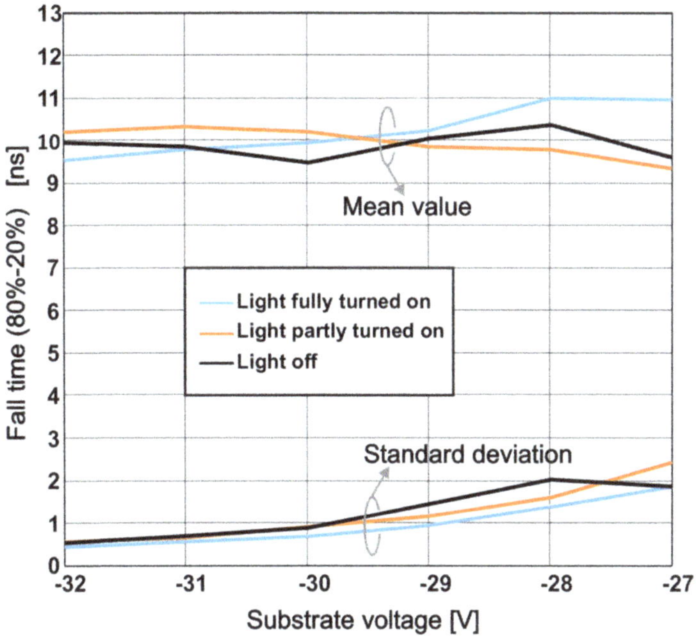

Figure 1.39. Mean values and standard deviations of the fall time of the PIN SPAD during self-quenching. © 2018 IEEE. Reprinted, with permission, from [52].

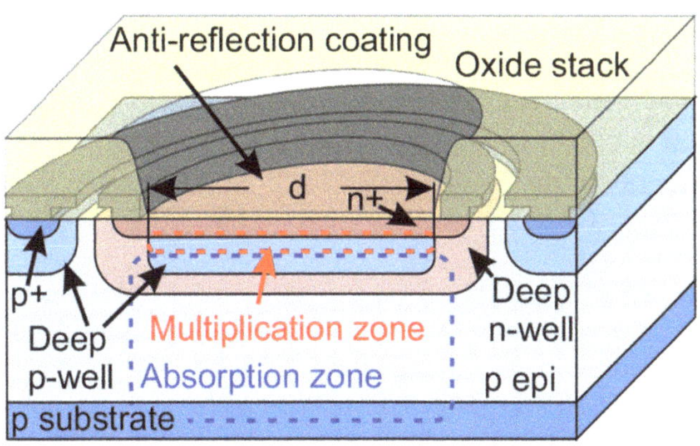

Figure 1.40. Structure of the thick SPAD with an antireflective coating in HV CMOS technology described in and reproduced from [55] with permission from SPIE.

The breakdown voltage of this SPAD was 68.25 V. The isolation capability of this HV CMOS process was 100 V. The HV CMOS process offered an antireflective coating, leading to a low-field (avalanche gain $M = 1$) responsivity of 0.41 A W^{-1} at 670 nm.

The spectral PDP of the HV CMOS SPAD is compared to several efficient SPADs from the literature in figure 1.41. At an excess bias of 3.5 V, its PDP was 22.1% at 785 nm. With an excess bias of 6.6 V, a PDP of about 45% was obtained at around 650 nm. The PDP was finally increased to 67.8% with an excess bias of 9.9 V

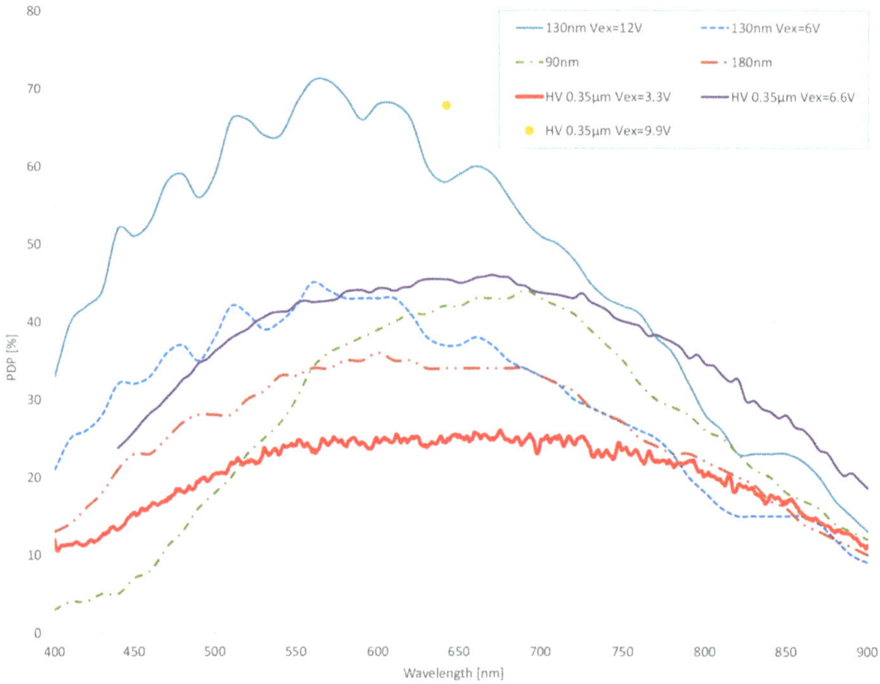

Figure 1.41. Comparison of the spectral photon detection probability of a 0.13 μm CMOS SPAD [44], a 90 nm CMOS SPAD [45], a 0.18 μm CMOS SPAD [46], and the thick HV CMOS SPAD in 0.35 μm technology.

at 642 nm [56]. The 0.35 μm HV CMOS SPAD achieves the highest PDPs with an excess bias voltage of 6.6 V from about 780nm onward. At 850 nm, the PDP of the 0.35 μm HV CMOS SPAD is about 28% with an excess bias of 6.6 V. At 642 nm and with a 9.9 V excess bias, the 0.35 μm HV CMOS SPAD reaches a higher PDP than the SPAD in 0.13 μm CMOS with a 12 V excess bias.

1.3.2.3 Comparison of avalanche transients for an HV SPAD and a PIN SPAD
The self-quenching experiments for the PIN SPAD (see figure 1.36–1.39) with the use of a gating circuit were extended to different SPAD diameters and also performed for an HV SPAD [54]. Several avalanche events, each triggered by a single photon, were overlaid using a storage oscilloscope, as shown, for instance, in figure 1.42 for a HV SPAD with a diameter of 48.2 μm at an anode voltage of −66 V. The light intensity of a halogen lamp was attenuated to obtain approximately equally distributed avalanche events (see the bottom part of figure 1.42).

Similar measurements were used to determine the fall times of the cathode voltage during self-quenching for PIN and HV SPADs with different diameters. Figure 1.43 shows these fall-time results (the 1/e decay time for these 80% to 20% fall times can be obtained by dividing these values by ln(4)). For SPADs with diameters larger than about 50 μm, the fall time reduces with increasing excess bias due to a larger avalanche current (see figure 1.46). The results show that SPADs with larger diameters quench themselves faster, i.e. they fire faster, and there is a trend for

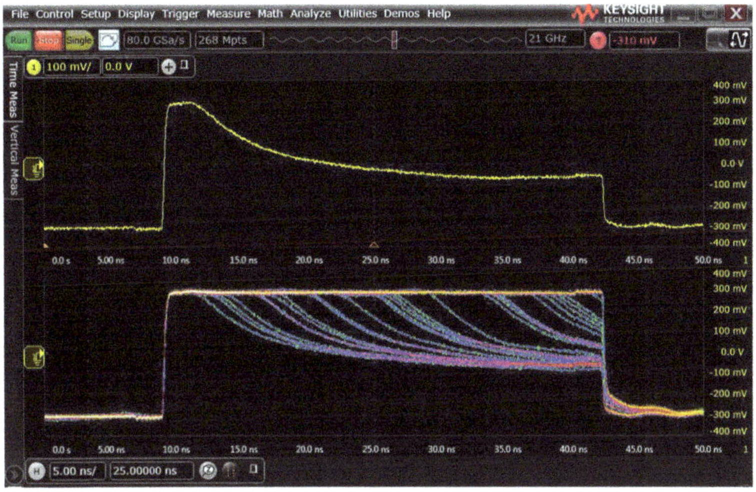

Figure 1.42. Oscilloscope screen shot of one avalanche (top) and several overlaid avalanches (bottom) showing the cathode voltage of an HV CMOS SPAD with a diameter of 48.2 μm. Reproduced from [54] with permission from MDPI.

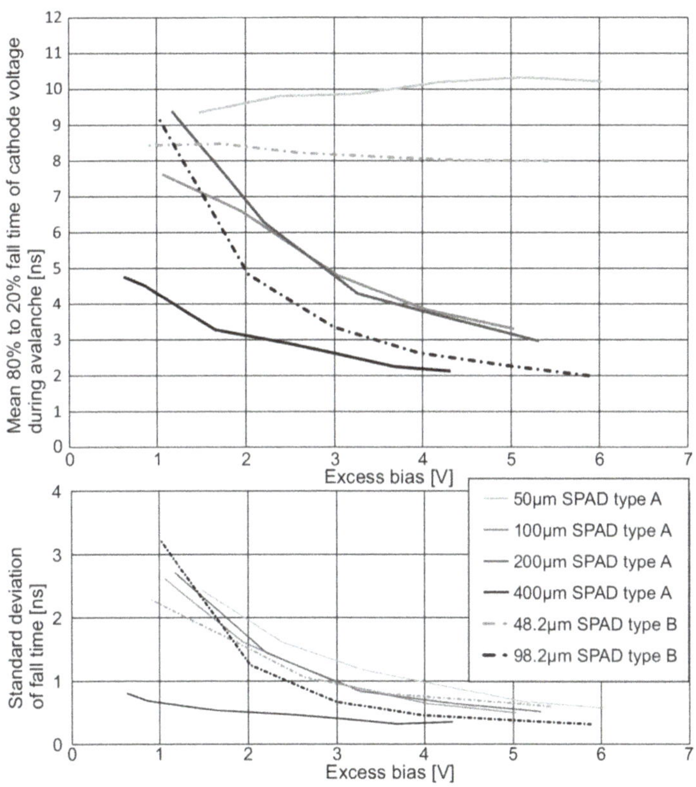

Figure 1.43. Fall times (80% to 20%) of the cathode voltages of a PIN SPAD (type A) and an HV CMOS SPAD (type B) for avalanche events. Reproduced from [54] with permission from MDPI.

the HV CMOS SPADs to fire faster than the PIN photodiode CMOS SPADs due to their larger avalanche currents. The smaller avalanche build-up time of the HV SPADs may be explained by their larger breakdown voltages and, in turn, by their higher electric field strengths.

The avalanche-current transients can be calculated from the voltage transients during self-quenching using $C_{Cat} \times dV_{Cat}/dt$, where C_{Cat} is the sum of the capacitance of the SPAD, the two bond pads, the gater input, and the GGB Industries picoprobe model 35 (50 fF). The first three were measured together using an Agilent 4284A precision LCR meter. For the PIN photodiode CMOS SPADs with diameters of 50, 100, 200, and 400 μm, the measured cathode node capacitances just below breakdown were 0.84, 1.12, 1.2, and 2.2 pF, respectively. For the HV CMOS SPADs with diameters of 48.2 and 98.2 μm, values of 0.88 and 1.12 pF, respectively, were measured at −40 V, which is the maximum voltage supported by the LCR meter. For the PIN SPAD with a 200 μm diameter, the transients obtained for the avalanche current for different anode voltages, i.e. different excess bias voltages, are depicted in figure 1.44. The transients of the avalanche current of the HV CMOS SPAD are shown in figure 1.45. In both figures, the avalanche current rises to a maximum during the first part of the discharge of the cathode node capacitance, which we call the avalanche build-up time. Subsequently, the excess bias decreases, the electric field strength in the multiplication zone decreases, and, in turn, the ionisation coefficients of the electrons and holes become smaller, which then leads to a lower avalanche current. When the discharge of the cathode node capacitance reaches the breakdown voltage level (i.e. the excess bias voltage becomes zero), the avalanche current vanishes. The maximum avalanche current

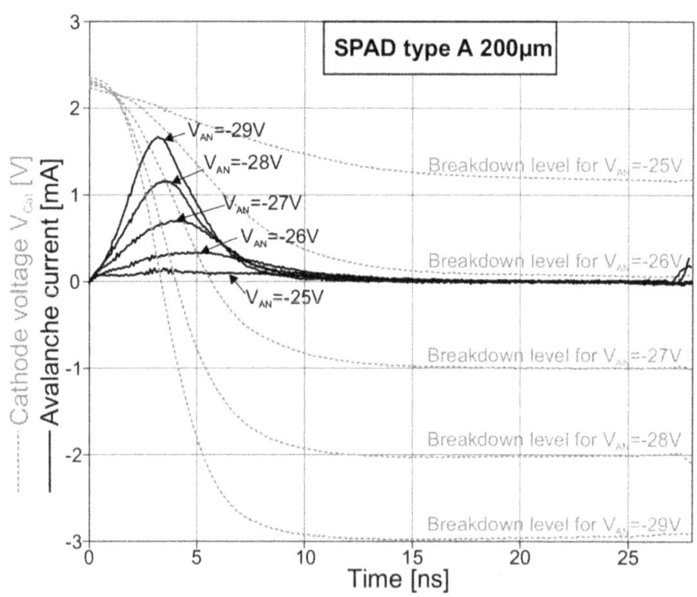

Figure 1.44. Avalanche-current transients for a PIN CMOS SPAD (type A) with a diameter of 200 μm for different anode voltages V_{AN} (the avalanche starts at 0 ns). Reproduced from [54] with permission from MDPI.

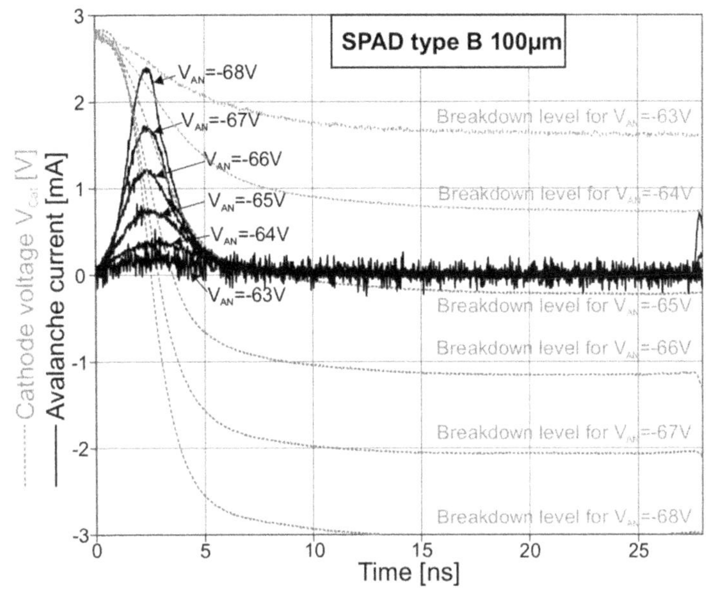

Figure 1.45. Avalanche-current transients for an HV CMOS SPAD (type B) with a diameter of 100 μm for different anode voltages V_{AN} (the avalanche starts at 0 ns). Reproduced from [54] with permission from MDPI.

increases with the excess bias voltage and with an increase in SPAD diameter for both types of SPAD. For the PIN photodiode CMOS SPAD, the maximum avalanche current is reached earlier for increasing levels of reverse bias, i.e. for increasing excess bias. For the HV CMOS SPAD, this effect is less pronounced, because the ratio of excess bias to breakdown voltage is considerably smaller for this type of SPAD than for the PIN photodiode CMOS SPAD.

The maximum avalanche currents of both types of SPAD are shown in figure 1.46. For both types of SPAD, the maximum avalanche current increases with the excess voltage and the SPAD diameter. This causes and explains the behavior shown in figure 1.43. Larger avalanche currents discharge the SPAD in a shorter time and reduce the fall times of the cathode node voltage during self-quenching. For the smallest SPAD diameters (about 50 μm) of both SPAD types, the bond-pad capacitances dominate and the fall times do not depend on the excess bias.

Obtaining the avalanche current from voltage transient measurements with the help of a gating circuit represents a new method of characterizing SPADs. The avalanche build-up time of SPADs can be measured in this way. The avalanche-current transients obtained for the PIN photodiode CMOS and the high-voltage CMOS SPADs during self-quenching reveal avalanche build-up times of approximately 3 ns and 2.5 ns, respectively. The avalanches in these SPADs last only a little more than 7 ns and 5 ns, respectively, which shows how fast active quenching circuits have to be to reduce the avalanche charge flowing through an SPAD and thereby the afterpulsing probability. Fortunately, these speed requirements have already been fulfilled with active quenchers in 0.35 μm CMOS, as demonstrated in [56–59] and summarized in subsections 3.1.2–3.1.4.

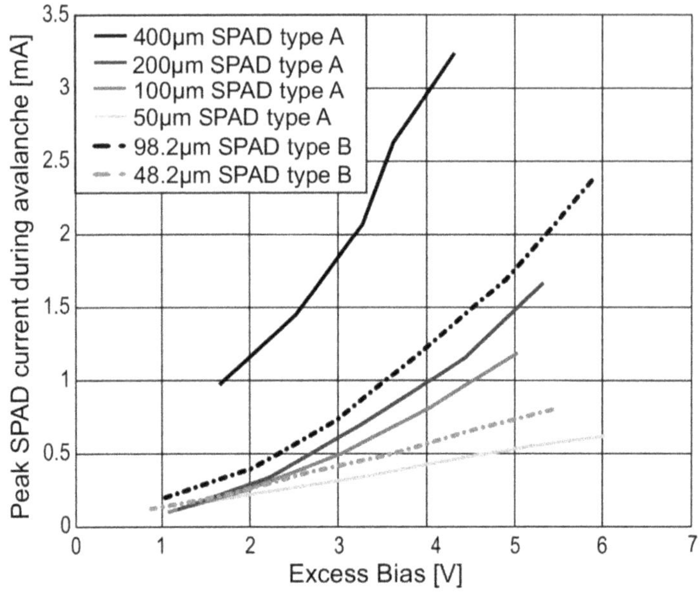

Figure 1.46. Dependence of peak avalanche current for a PIN CMOS SPAD (type A) and an HV CMOS SPAD (type B) on excess bias. Reproduced from (54) with permission from MDPI.

1.3.2.4 Thick modulation-doped SPAD in HV CMOS

Modulation doping has been used in linear-mode avalanche photodiodes to increase the bandwidth [60]. This motivated us to also investigate the influence of modulation doping on the performance of SPADs. Modulation doping is a pure design measure that reduces the effective doping in the multiplication zone. An important advantage is that modulation doping does not need any process modifications. Modulation doping is performed by implementing a hole pattern in the mask layer, which defines the implantation of the p-well or deep p-well that forms the multiplication region. As a consequence, the total dose and therefore the effective doping of the multiplication layer are reduced. Due to the thermal budget of the CMOS process, the holes present after implantation close because of dopant diffusion. It is, however, clear that the multiplication region will not be perfectly homogenous when this modulation doping is applied. The higher the thermal process budget, the better the homogeneity. Therefore, modulation doping works better in high-voltage CMOS technologies than in standard (digital) CMOS technologies.

The influence of modulation doping on the performance of a thick HV CMOS SPAD was investigated in [61]. Figure 1.47 compares the cross section of this modulation-doped SPAD (SPAD2) to that of a simple thick HV CMOS SPAD (SPAD1).

A modulation-doping factor of about 90% was used. This was achieved by using a specific hole pattern for the DPW, i.e. for SPAD1. Holes were drawn inside the layout mask of this DPW, as depicted in the bottom part of figure 1.47, to prevent boron implantation within these holes. The hole diameter was 0.9 μm and the gap

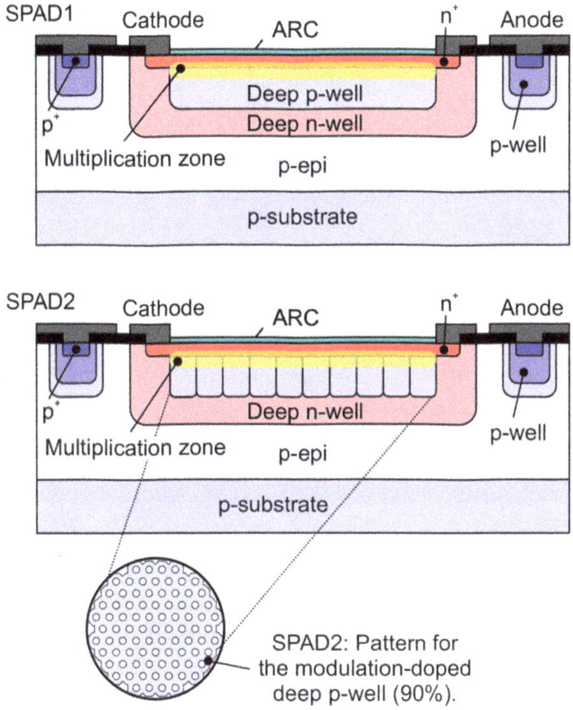

Figure 1.47. Cross sections of a thick SPAD (top: SPAD1) and a thick modulation-doped SPAD as well as a top view of the modulation-doping hole pattern (bottom: SPAD2) in a 0.35 μm CMOS SPAD. Reproduced from [61] with permission from SPIE.

between holes was 4.3 μm. However, in order for this to work properly, the dimensions of the holes and the gaps between the holes need to stay below a technology-dependent limit, which depends on the thermal budget of the process.

The active diameters (i.e. the DPW diameter) of both SPADs were 85 μm. Both SPADs were implemented together with the same cascoded active quencher. A chip photo of one of the two test chips is presented in figure 1.48. The dead time of the active quenching circuit used for both SPADs was tuneable from 5.8 ns to 33.4 ns. The total chip areas of the realized OEICs were 680 × 980 μm^2.

Because the important parameters of SPADs, such as the breakdown voltage, the dark count rate, and the afterpulsing probability fluctuate, depending on the device's position on the wafer, three pairs of samples were taken for SPAD1 and SPAD2 at the wafer borders (left, top, right), and three pairs of samples were taken from the center of a wafer.

These devices were mounted in a dark box on a thermoelectric cooler that regulated them to 25 °C. The outputs of the active quenching circuit were connected to a National Instruments NI-5162 digitizer. A recording time of 10 s was used for each bias point. The dependencies of the DCR, APP, and PDP on the excess bias voltage (V_{ex}) were measured. The APP was reported for dead times of t_d=5.8 ns and t_d=33.4 ns; the DCR was only reported for the longer dead time. The DCR and APP results for SPAD1 are depicted in figures 1.49(a) and (c), respectively.

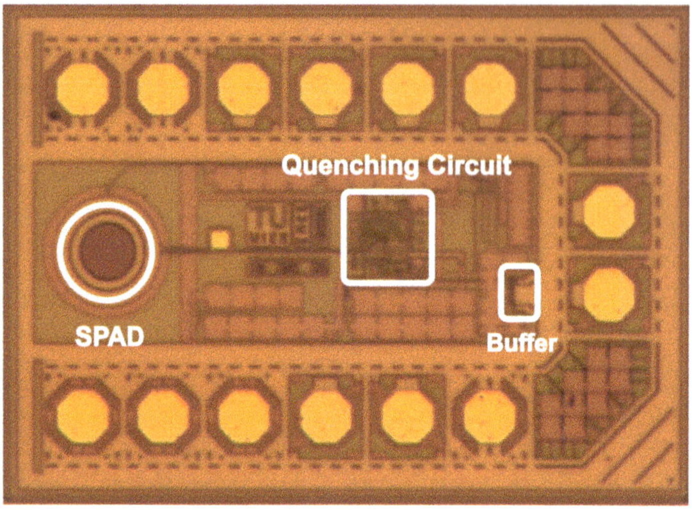

Figure 1.48. Microphotograph of the test chip used for characterization of thick SPADs with and without modulation doping in 0.35 μm HV CMOS. Reproduced from [61] with permission from SPIE.

The breakdown voltage of SPAD1 varied from 65.5 V to 70.2 V for the chosen samples. Considering only the SPADs from the wafer's center (SPAD1-center3, dotted lines), the variation of the breakdown voltage was about 0.4 V. The lowest DCRs were obtained for the samples from the center. However, one DCR curve of the center SPADs was in the same range as those of the SPADs from the wafer's periphery. The ratio between the highest (140.4 kcps) and lowest (28.8 kcps) DCRs at $V_{ex} = 6.6$V was about 4.9. Also, the APP showed a large fluctuation between samples (figure 1.49(c)). The APPs varied by factors of 1.7 and 3.2 at dead times of $t_d = 5.8$ ns and $t_d = 33.4$ ns, respectively. In contrast to the DCR, SPADs from the periphery of the wafer tended to have better APPs.

The performance of SPAD2 is shown in figure 1.49(b) and (d). As mentioned above, SPAD2 used a modulation-doped DPW. This modulation technique generated a well with a reduced effective dopant level and therefore increased the breakdown voltage. For the samples described here, the breakdown voltage varied from 80.1 V to 85.5 V. The breakdown voltage spread of the center SPADs was about 0.6 V. It is clearly visible that the DCR is improved by about a factor of two compared to SPAD1 for corresponding excess bias voltages. In addition, the APP is lower for SPAD2, as depicted in figure 1.49(d). The reduced DCR and APP may be explained by a lower effective excess bias voltage V_{ex} (i.e. the smaller ratio of the excess bias voltage to the breakdown voltage) [62, 63].

We next discuss the PDP. The PDP was corrected for the DCR and the APP. The DCR and the APP depend more strongly than the PDP on the position of the SPAD on the wafer. Because the 0.35 μm high-voltage CMOS process used for this comparison is a mature process, a low non-uniformity of the PDP over the wafer is to be expected, as described, for example, in [64]. Therefore, the PDP was only measured for one sample per SPAD structure. The PDP of an SPAD that had the

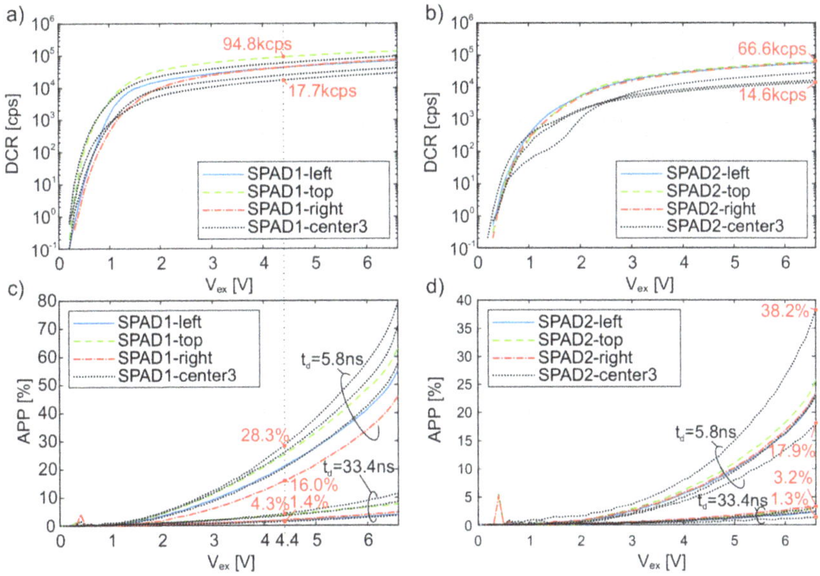

Figure 1.49. Dependence of the dark count rate on the excess bias for a dead time t_d of 33.4 ns: (a) SPAD1 and (b) SPAD2. Dependence of the afterpulsing probability on the excess bias: (c) SPAD1 and (d) SPAD2. Reproduced from [61] with permission from SPIE.

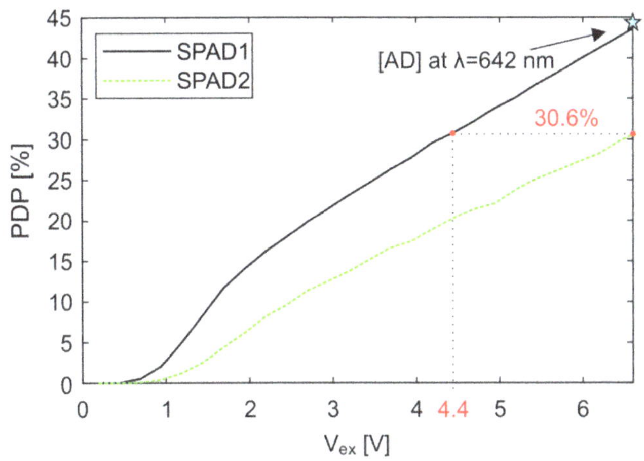

Figure 1.50. Dependence of the photon detection probabilities of SPAD1 and SPAD2 in 0.35 μm HV CMOS on the excess bias for a wavelength of 642 nm. Reproduced from [61] with permission from SPIE. [AD]: [56].

same structure as SPAD1 [56] showed a PDP of 44% for λ=642nm at a 6.6 V excess bias, which almost perfectly matched the PDP presented in [61] and shown here in figure 1.51. According to this figure, the PDP indicates the disadvantage of SPAD2. Because of the higher breakdown voltage and, in turn, its lower effective excess bias, the PDP of SPAD2 is reduced compared to that of SPAD1. The maximum PDP of SPAD1 is 43.6%, in contrast to 30.6% for SPAD2 (both at the largest excess bias of 6.6 V). Therefore, a direct comparison of SPAD1 and SPAD2 was performed for the

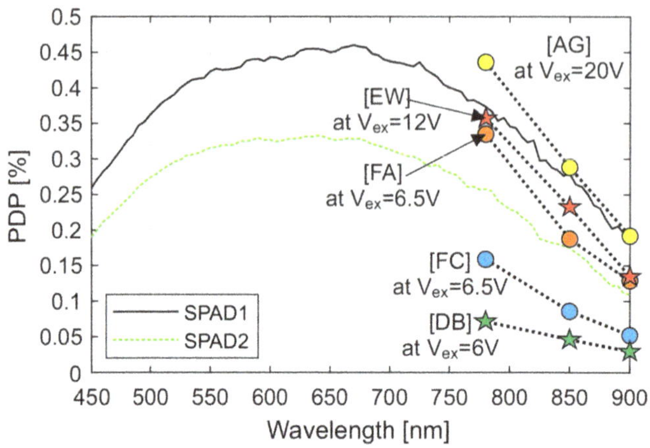

Figure 1.51. Dependence of the photon detection probabilities of SPAD1 and SPAD2 in 0.35 μm HV CMOS on wavelength at an excess bias of 6.6 V compared to published results. Reproduced from [61] with permission from SPIE. ([AG]: [65]), [DB]: [66], [EW]: [44], [FA]: [67], [FC]: [68].

same PDP. SPAD1 shows a PDP of 30.6% at an excess bias of about 4.4 V (see figure 1.50). The values of the DCR and the APP for this excess bias are highlighted in figure 1.49(a) and (c), respectively. The afterpulsing probability and dark count rate of SPAD2 are still somewhat better (the APP of SPAD2 is 1.3% to 3.2% at V_{ex}=6.6 V, in comparison to the APP of SPAD1, which is 1.4% to 4.3% at V_{ex}=4.4 V; the DCR of SPAD2 is 14.6 kcps to 66.6 kcps at V_{ex}=6.6 V in comparison to that of SPAD1, which is 17.7 kcps to 94.8 kcps at V_{ex}=4.4 V) for a dead time of 33.4 ns. Only one sample of SPAD2 from the wafer's center shows a worse APP (38.2% at V_{ex}=6.6 V in comparison to 28.3% at V_{ex}=4.4 V for SPAD1) for a dead time of 5.8 ns.

As verified in [69], the modulation-doping technique increases the breakdown voltage and therefore strengthens the electric field in the thick absorption zone. Additionally, the depleted region expands deeper towards the substrate, promoting the PDP at long wavelengths. Therefore, the spectral PDP distribution is also shown in figure 1.51 for an excess bias of 6.6 V for SPAD1 and SPAD2. At wavelengths of 780nm, 850 nm, and 900 nm, SPAD1 achieves PDPs of 37.4%, 27.9%, and 18.6%, respectively. The PDPs of SPAD2 for the same excess bias are 25.7%, 17.5%, and 10.9%, respectively. The maximum PDP of 46.0% of SPAD1 is present at λ = 670 nm. For SPAD2, the maximum PDP of 33.2% is located at 640 nm. For these NIR wavelengths in particular, both SPAD structures achieve outstanding PDP results in comparison to other integrated CMOS SPADs [44, 66] and about the same PDPs as those of SPADs produced in dedicated custom processes [65, 67, 68]. The PDP values of the integrated CMOS SPADs presented in [44, 66] at 780nm, 850nm, and 900nm are represented as stars in figure 1.51, and the PDPs of the SPADs introduced in [65, 67, 68], which used optimized custom processes are shown as circles. Please note that for the PDPs of the SPADs of [44] and [65], much higher excess bias voltages were applied, compared to those of the other SPADs. The PDPs of SPAD1 and SPAD2 can be further raised by increasing the excess bias. The PDPs for wavelengths of 780nm, 850 nm, and 900 nm, as well as the excess bias voltage of SPAD1 and SPAD2 are summarized and compared with those described in references [44] and [65–68] in table 1.1

The performances of different detectors are often compared using the noise-equivalent power (NEP). The NEP of SPADs was defined as follows in [70]:

$$NEP = \frac{hc}{\lambda\sqrt{2DCR/PDP}}, \tag{1.30}$$

where h is Planck's constant, c is the vacuum velocity of light, and λ is the wavelength of the incident light. The smaller the value, the better the performance of the SPAD. Table 1.2 lists the NEPs of SPAD1 and SPAD2 for wavelengths of 780 nm, 850 nm, and 900 nm. For both detectors, the SPAD with the smallest DCR was chosen. The NEP of SPAD1 is slightly better than that of SPAD2. Nevertheless, the difference is small and it is easy to tune the NEP by changing the active area or the temperature of the SPAD, since these measures mainly influence the DCR, while the PDP is almost unaffected.

We now summarize and compare SPAD1 with the standard wells and SPAD2 with the modulation-doped deep p-well. SPAD1 has the original deep p-well of the HV CMOS process and exhibits very high PDP values of 27.9% at 850 nm and 18.6% at 900 nm. SPAD2's modulation doping was implemented in order to reduce the effective doping concentration and in turn to increase its breakdown voltage. This enhances the electric field inside the absorption region and in addition creates a thicker space-charge region. In the linear mode, in which the device is exploited as an avalanche photodiode (APD), this increases the bandwidth of the device. The results

Table 1.2. PDP comparison with the state of the art [61].

SPAD	V_{ex}[V]	Technology	PDP at 780 nm [%]	PDP at 850 nm [%]	PDP at 900 nm [%]
SPAD1	6.6	0.35 μm HV CMOS	37.4	27.9	18.6
SPAD2	6.6	0.35 μm HV CMOS	25.7	17.5	10.9
[44]	12	130 nm CMOS	35.8	23.4	13.6
[66]	6	0.35 μm CMOS	7.3	4.8	2.7
[68]	6.5	Custom	15.9	8.6	5.2
[65]	20	Custom	43.6	28.8	19.2
[67]	6.5	Custom	33.5	18.8	12.9

Table 1.3. Noise-equivalent powers (NEPs) of SPAD1 and SPAD2 at V_{ex}=6.6 V [61].

SPAD	V_{ex} [V]	NEP at 780 nm [aW/$\sqrt{(Hz)}$]	NEP at 850 nm [aW/$\sqrt{(Hz)}$]	NEP at 900 nm [aW/$\sqrt{(Hz)}$]
SPAD1	6.6	78.3	90.7	111.1
SPAD2	6.6	85.8	104.0	131.8

above show that modulation doping is an effective way to vary the breakdown voltage of a certain SPAD structure. SPAD2, which has modulation doping, possesses reduced DCR and APP compared to SPAD1, which has no modulation doping. The DCR and APP (for a dead time of 33.4 ns) of SPAD2 with modulation doping are, on average, about 27% and 20% lower than the corresponding values for SPAD1 for the same photon detection probability of 30.6%. However, as a drawback, the increased breakdown voltage reduces the effective excess voltage V_{ex} and therefore causes a lower PDP for the same applied excess voltage. Nevertheless, the PDP values presented for NIR wavelengths for the modulation-doped SPAD2 of 17.5% at 850 nm and 10.9% at 900 nm are in the same range as those reported for state-of-the-art SPADs (integrated into CMOS technologies and in custom processes).

1.4 A model for photon detection probability

Since the PDP of so-called SPADs is often much smaller than 100%, and since the PDP depends strongly on the wavelength and the structure of the SPAD, it is of great interest to develop a comprehensive model that predicts the PDPs of SPADs in CMOS and BiCMOS technologies, i.e. one that considers doping as well as the isolation and passivation stack. During the development of such a model, it was actually found that the widely accepted Lambert–Beer law, which is implemented in off-the-shelf technology computer-aided design (TCAD) device simulation programs used to calculate light intensity and photogeneration in silicon device regions, is not accurate enough to describe the measured PDP spectra [71]. In the following, a precise method will be described that can calculate the PDPs of SPADs, not only in opto-application-specific integrated circuit (opto-ASIC) processes with antireflective coatings, but also in (Bi)CMOS technologies that do not offer an antireflective coating, i.e. using their standard isolation and passivation stack. Figure 1.52 shows the cross section of such an SPAD structure in HV CMOS. On top of the active region of the SPAD, there is an isolation stack consisting of several deposited oxide layers (intermetal dielectrics) and a covering passivation layer composed of silicon

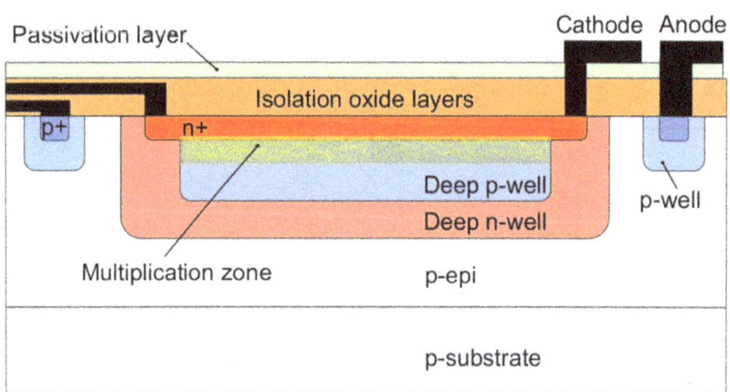

Figure 1.52. Cross section of an HV CMOS SPAD with an isolation and passivation stack.

nitride or oxynitride. This oxide and passivation stack will become important below for the new PDP model.

The doping structure of this device is the same as those described above. Figure 1.53 shows the electric field within this SPAD, which exceeds the critical electric field strength required for impact ionization. We clearly see that the SPAD works in area breakdown. The electric field is necessary to model the impact ionization rates of electrons and holes. This figure also shows the boundaries of the depletion region, which define the widths of the neutral regions between the silicon surface and the upper depletion boundary as well as between the lower depletion boundary and the p substrate.

The next quantity we need for the model is the light intensity in the silicon; this allows us to obtain the distribution of photogenerated electron–hole pairs. The first attempt to calculate this was based on a standing-wave model for the isolation and passivation stack using the effective index of refraction of the effective thickness of one dielectric layer. Figure 1.54 shows the transmission of the isolation and

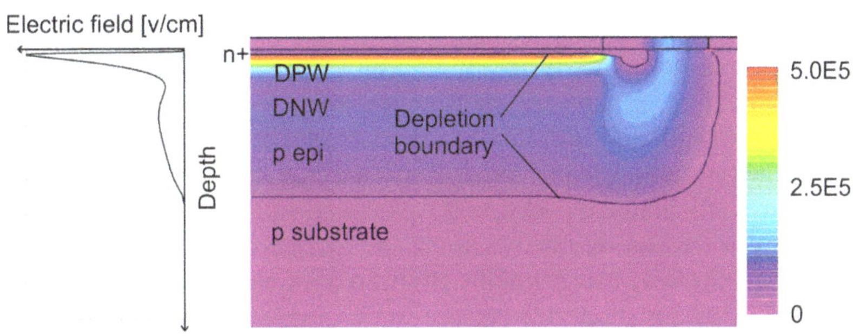

Figure 1.53. Simulated electric field in the HV CMOS SPAD (V_{ex} = 6.6 V). © 2020 IEEE. Reprinted, with permission, from [71].

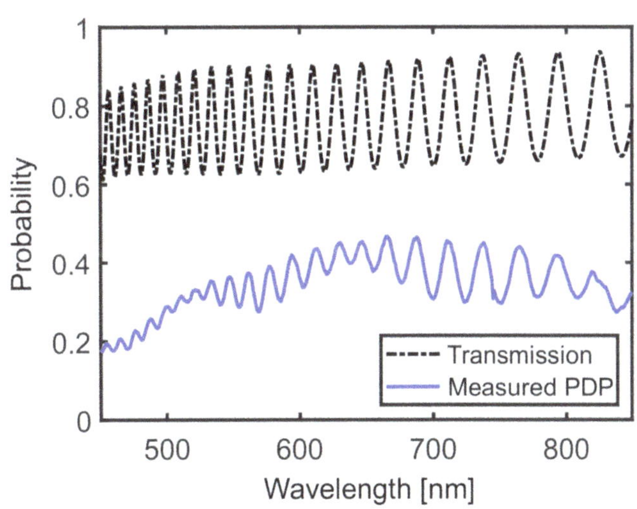

Figure 1.54. Transmission in the isolation and passivation stack, assuming a standing wave.

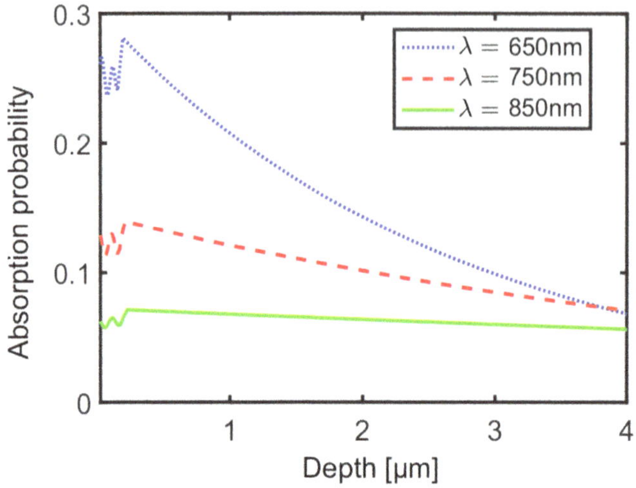

Figure 1.55. Photon absorption probability, assuming a standing wave in the isolation and passivation stack.

passivation stack in comparison to the measured PDP. This simple model of one standing wave in the isolation and passivation stack allows us to obtain the exact locations of the maxima and minima in the PDP spectrum. With this one-standing-wave model, the photon absorption distribution in the silicon device region depicted in figure 1.55 was calculated by solving the Maxwell equations numerically using the transmission from the standing-wave approach for wavelengths of 650 nm, 750 nm, and 850 nm. The photon absorption distribution obtained deviates remarkably from the Lambert–Beer exponential decay close to the surface, which makes it difficult to describe quantum efficiency and PDP correctly for light with short wavelengths, i.e. in the blue and green parts of the spectrum.

The photon detection probability is a function of the photon absorption probability ($P_{ab}(\lambda, x)$) and the total avalanche-triggering probability ($P_{av}(x)$) and can be obtained as follows [72]:

$$\mathrm{PDP}(\lambda) = \int_0^{\infty} \mathrm{P}_{ab}(\lambda, x) \times \mathrm{P}_{av}(x) dx, \tag{1.31}$$

where the upper integration boundary should be understood in the sense that all incident photons, therefore, all photogenerated carriers are considered. In practice, it is sufficient to use $5/\alpha$ instead of infinity to cover more than 99% of the photo-generated carriers.

The widely used photon absorption probability depends on the wavelength λ and the absorption depth x, according to the Lambert–Beer law:

$$P_{ab}(\lambda, x) = \alpha(\lambda)e^{-\alpha(\lambda)x}, \tag{1.32}$$

where $\alpha(\lambda)$ is the dependence of the optical absorption coefficient on λ. However, the solution of Maxwell's equations leads to a deviation from this exponentially decaying dependence below the silicon surface.

To calculate the PDP, we also need the avalanche-triggering probability in addition to the photon absorption distribution. To calculate the probability that a self-sustaining avalanche will be triggered by an electron or by a hole photogenerated at a depth x within the depleted region, the following coupled equations have to be solved [73]:

$$\begin{aligned} \frac{\partial P_e}{\partial x} &= (1 - P_e)\gamma_e(P_e + P_h - P_e P_h), \\ \frac{\partial P_h}{\partial x} &= (1 - P_h)\gamma_h(P_e + P_h - P_e P_h). \end{aligned} \tag{1.33}$$

where γ_e and γ_h are the impact ionization coefficients of electrons and holes, respectively.

The avalanche-triggering probability suggested in [74, 75] was modified with respect to diffusing carriers from the neutral regions in [71]. The probability of diffusion (P_{diff}) of a photogenerated minority carrier through the neutral regions into the depletion region is given by [71]:

$$P_{diff}(x) = \begin{cases} e^{-(\frac{w_1 - x}{L_h})} & \text{for } x < w_1, \text{ neutral n } - \text{ type region (above)}, \\ e^{-(\frac{x - w_2}{L_e})} & \text{for } x > w_2, \text{ neutral p } - \text{ type region (below)}, \end{cases} \tag{1.34}$$

where L_h and L_e are the diffusion lengths of the holes and electrons, respectively. Here, w_1 and w_2 correspond to the top and bottom boundaries between the neutral regions and the depletion region. Within the depleted region, P_{diff} is equal to one.

The total avalanche probability ($P_{av}(x)$) when a photon is absorbed at x is obtained by considering the probability that either an electron or a hole triggers an avalanche, which is given by:

$$P_{av}(x) = [P_e(x) + P_h(x) - P_e(x)P_h(x)] \times P_{diff}(x). \tag{1.35}$$

The avalanche-triggering probabilities P_e and P_h for electrons and holes, respectively, are shown in figure 1.56. In the depleted region below the multiplication zone (values of x between 0.8 and 10 μm), $P_e(x)$ is constant, because electrons photogenerated in this range cross the whole multiplication zone and all of them have the same chance of starting a self-sustaining avalanche. However, if an electron is photogenerated above the multiplication zone, it reaches the n+ cathode without entering the multiplication zone, with the result that P_e is zero close to the silicon surface. For electrons photogenerated within the multiplication zone (values of x between 0.2 and 0.8 μm), P_e increases from 0 to the maximum value. Since holes are transferred to the anode, P_h behaves in the reverse manner, and because the hole impact ionization coefficient is smaller than the electron impact ionization coefficient in silicon, the maximum value of P_h is smaller than that of P_e. For charge carriers photogenerated above and below the depleted regions, the probability of diffusion into the depleted region is smaller than one and this is accounted for by P_{diff} (equations (1.34) and (1.35)). These diffusing carriers contribute to the maximum values of P_e and P_h.

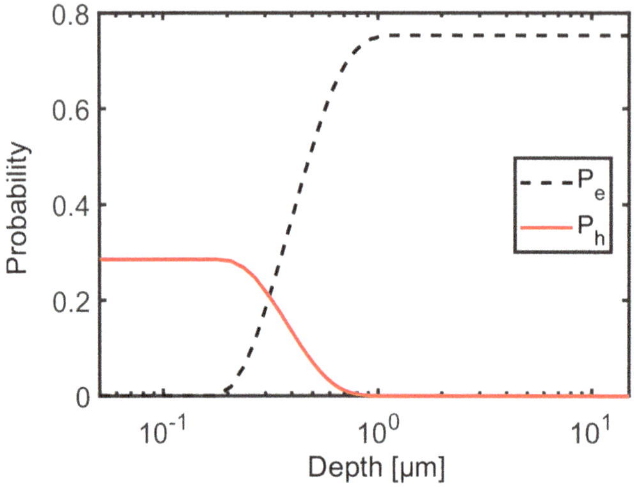

Figure 1.56. Avalanche-triggering probabilities for electrons and holes as a function of absorption depth. © 2020 IEEE. Reprinted, with permission, from [71].

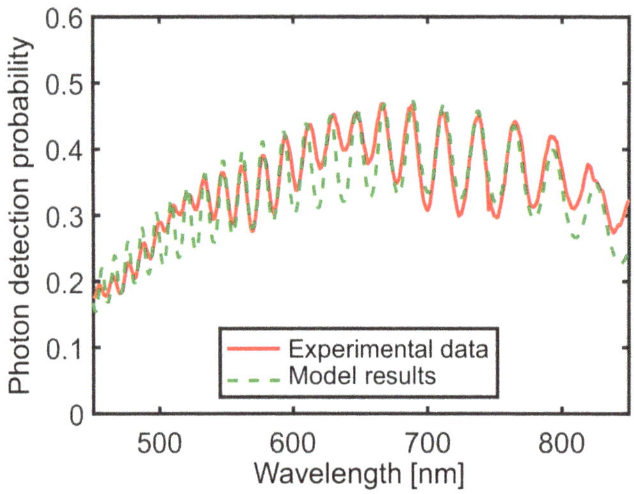

Figure 1.57. Measured and simulated PDP spectra at $V_{ex} = 6.6$ V, assuming one standing wave in the oxidation and passivation stack.

We now have everything required to solve equation (1.31). The results for the modeled PDP are compared to the PDP shown in figure 1.57. The electron and hole avalanche-triggering probabilities, the electric field, and the boundaries of the depleted zone were simulated using the Geiger-mode feature of the ATLAS simulator [6] by considering the photon transmission (Figure 1.54) and absorption profile and assuming one standing wave (figure 1.55). The parameters used in ATLAS are listed in table 1.4. The locations of the maxima and minima are

Table 1.4. Parameters used in the ATLAS TCAD simulations [76].

Parameter	Description	Value
a_n	Impact ionization parameters of electrons [77]	7.03×10^5 1/cm
E_{n_crit}		1.231×10^6 V/cm
a_p	Impact ionization parameters of holes [77]	1.58×10^6 1/cm
E_{p_crit}		2.036×10^6 V/cm
V_{br}	Breakdown voltage	25 V
τ_{n}	Electron lifetime	200 μ s
τ_{p}	Hole lifetime	200 μs
L_{n}	Electron diffusion length	270 μm
L_{p}	Hole diffusion length	90 μm
w_1	Top boundary of the depleted region	220 nm
w_2	Bottom boundary of the depleted region	11.96 μm

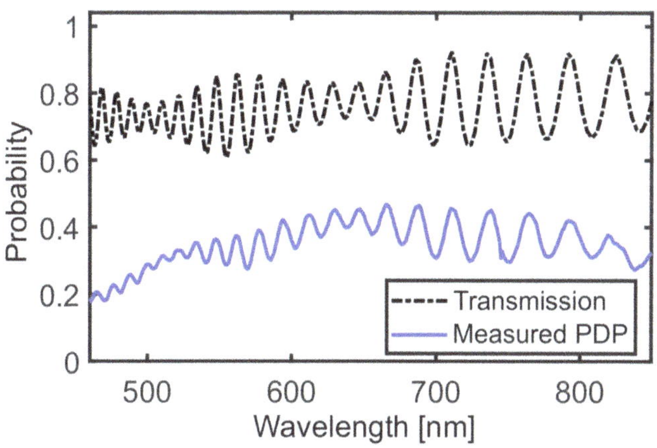

Figure 1.58. Transmission in the isolation and passivation layers, assuming two standing waves. © 2020 IEEE. Reprinted, with permission, from [71].

reproduced well by the model, but between about 500 and 550 nm as well as between 610 and 660 nm, there is a large deviation from the measured PDP values.

To improve the modeled PDP values in these spectral ranges, two standing waves were assumed, one in the oxide layers of the isolation stack and one in the passivation layer composed of silicon nitride. Figure 1.58 shows that the two standing waves modulate each other and that the locations of the maxima and minima in the modeled transmission are a good fit for those of the measured PDP. The solution of the Maxwell equations again results in a deviation from an exponential decay of the photogeneration inside the silicon (see figure 1.59). This deviation from the exponential decay is caused by the penetration of the standing wave from the isolation layers into the silicon [76].

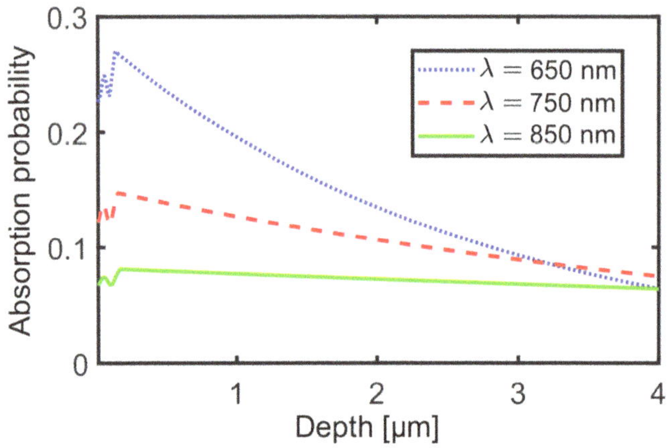

Figure 1.59. Photon absorption probability, assuming two standing waves in the oxide and passivation layers. © 2020 IEEE. Reprinted, with permission, from [71].

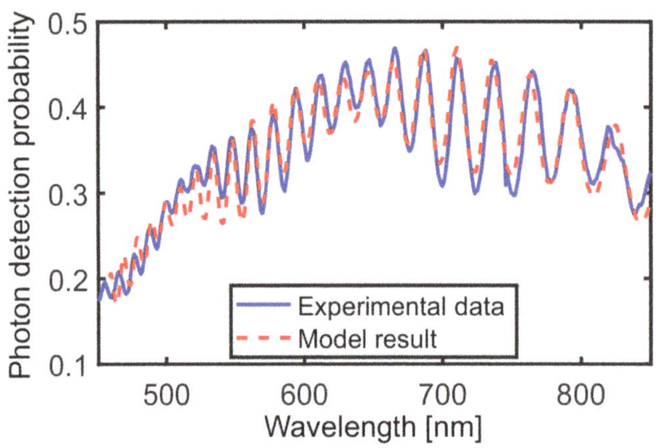

Figure 1.60. Measured and simulated PDP spectra of the thick high-voltage CMOS SPAD at V_{ex} = 6.6 V, assuming two standing waves in the oxide and passivation layers. © 2020 IEEE. Reprinted, with permission, from [71].

Finally, we can compare the PDP spectrum obtained using the two-standing-wave approach with the measured PDP spectrum in figure 1.60. There is now an almost perfect agreement in the red spectral range and also a better fit in the blue/green range. The remaining small deviation may be due to slight differences in the optical index of refraction within the different deposited oxide layers or due to inhomogeneous layer thicknesses within the light-sensitive area of the SPAD.

The PDP model developed was also applied to a PIN photodiode CMOS SPAD [76]. The structure of the PIN photodiode CMOS SPAD is shown in figure 1.61. The incident angle of the light θ_0 and the corresponding angle in the silicon region θ_{Si} are also defined in this figure.

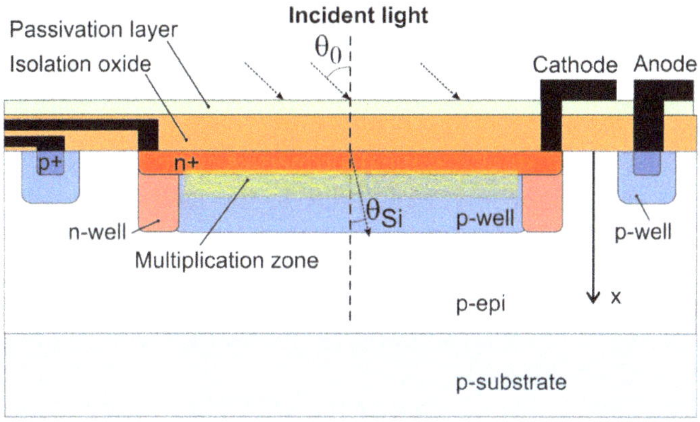

Figure 1.61. Cross section of the thick PIN photodiode CMOS SPAD with oxide and passivation layers [76].

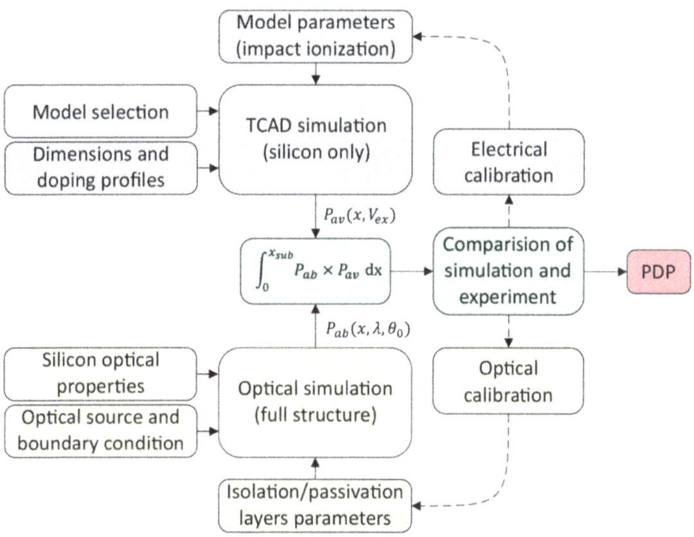

Figure 1.62. Flow chart of the PDP modeling method developed [76].

The methodology of the PDP modeling procedure is depicted in figure 1.62. The avalanche probability P_{av} is obtained from an ATLAS TCAD device simulation, and P_{ab} is obtained by optical simulation. The optical simulation consists of calculating the transmission in the oxide and passivation stack, as well as the optical power decay (i.e. photon absorption in the silicon region), by applying electromagnetic simulation through solving the Maxwell equations. The upper integration boundary x_{sub} represents the thickness of the silicon region, which is typically a few hundred micrometers. In practice, for many wavelengths, it will be sufficient to use $5/\alpha$ instead of x_{sub} to reduce the error to less than 1% (the condition $5/\alpha < x_{sub}$ has to be fulfilled, of course). In the device simulation, we used the SPAD geometry from the layout, information from the process development kit, and the doping profiles, which were made available by the ASIC foundry. The necessary models for impact ionization, generation, recombination, and mobility were applied. The impact ionization parameters had to be calibrated to reproduce the measured PDP data. Definition of the light source, implementation of the known optical properties (e.g. the refractive indices) of silicon, oxide, nitride, and air, and calibration of the isolation and passivation stack's optical properties were necessary for the optical simulation. This calibration was needed due to the lack of available information regarding the exact oxide and passivation layer thicknesses and their refractive indices.

The modeled PDP is compared to the measured PDP spectra for two excess bias values in figure 1.63. The figure shows very good agreement between the modeled results and the measured results.

Table 1.4 lists the parameters used in the TCAD device simulations performed using ATLAS. The same set of parameters was used for the PDP calculations of the HV CMOS SPAD and the PIN photodiode CMOS SPAD.

The results described above corroborate a very good understanding of the physics involved in avalanche events. Figure 1.64 depicts the photon absorption probabilities for a wide wavelength range in the different regions of the SPAD. A large portion of the photons with wavelengths shorter than about 500 nm is absorbed in the upper neutral region. Photons with longer wavelengths are mainly absorbed in the depleted region of the SPAD. The PDP for shorter wavelengths is determined by holes generated near the surface (see figure 1.56), and a lower PDP is expected because the avalanche-triggering probability of holes is about 2.5 times smaller than that of electrons (see figure 1.56).

The very good dependence of the modeled PDP on the excess bias voltage above a threshold of approximately 2 V is visible in figure 1.65 for four distinct wavelengths. For excess bias values below this threshold, the readout sensitivity of the active quencher affects the measured PDP. This effect is not included in the described

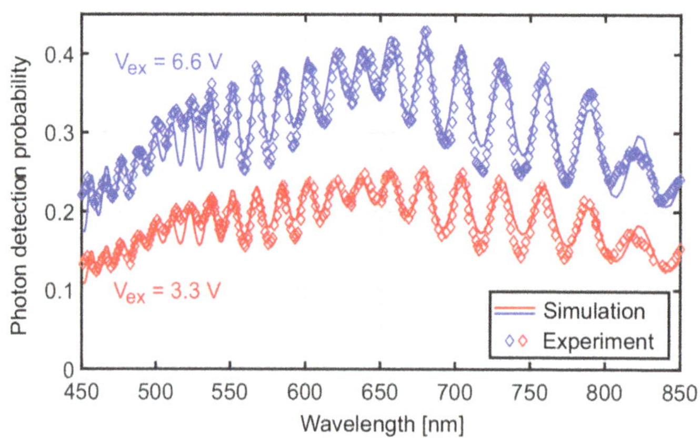

Figure 1.63. Measured and simulated PDP spectra of the thick PIN photodiode CMOS SPAD for V_{ex} = 3.3 V and V_{ex} = 6.6 V, assuming two standing waves in the oxidation and passivation layers. © 2021 IEEE. Reprinted, with permission, from [76].

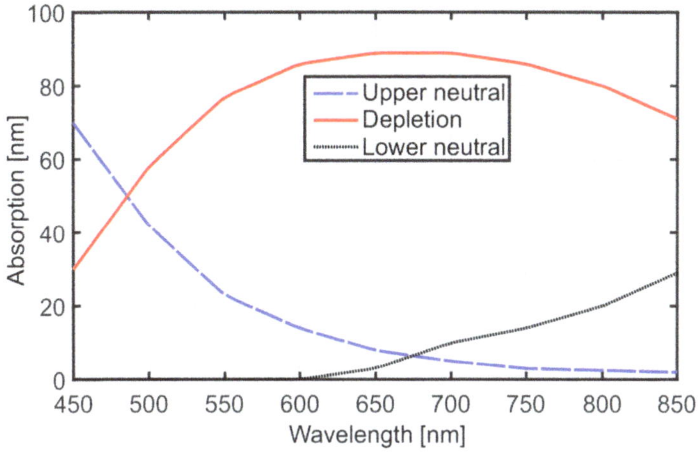

Figure 1.64. Wavelength dependence of the contributions of the upper neutral region, the depletion region, and the lower neutral region of the PIN photodiode CMOS SPAD on photon absorption. © 2021 IEEE. Reprinted, with permission, from [76].

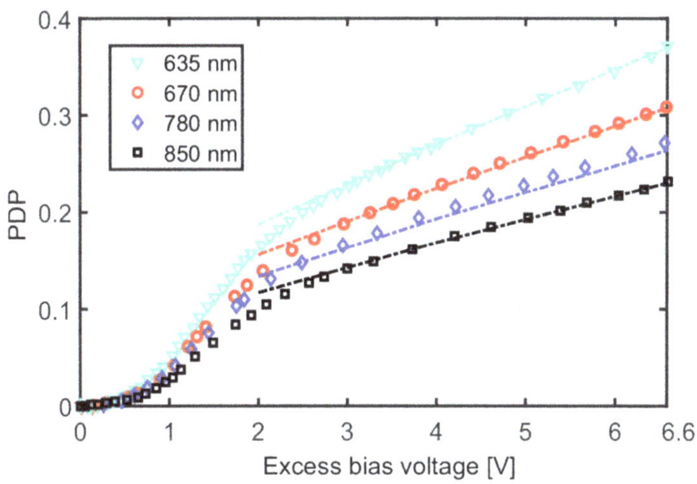

Figure 1.65. Comparison of the measured (symbols) and modeled (lines) PDPs of the PIN photodiode CMOS SPAD in dependence on the excess bias voltage for four different wavelengths. © 2021 IEEE. Reprinted, with permission, from [76].

modeling and simulation method. The linear increase of the PDP with excess bias stems from the linear increase of the electric field strength in the multiplication zone with the voltage. This follows from a negligible variation of the thickness of the depleted region for voltages above breakdown, as verified by device simulations for the PIN photodiode CMOS SPAD [76]. Figure 1.65 exhibits the different slopes of the PDP for the four wavelengths. However, when the PDP curves are normalized to their corresponding value at 7 V, independence from the wavelength follows (see figure 1.66).

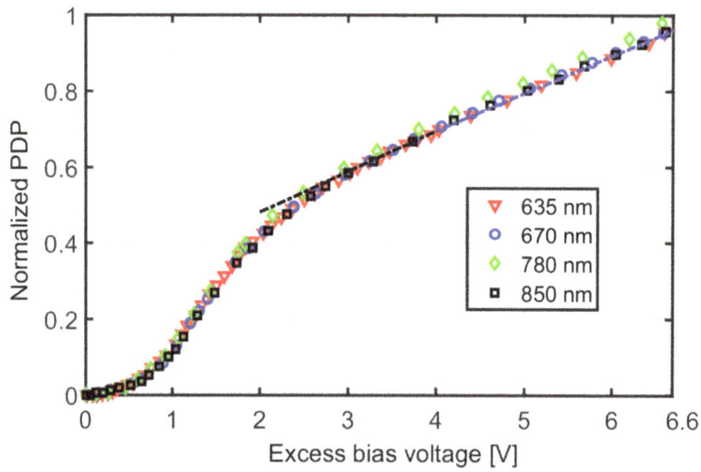

Figure 1.66. Comparison of the measured (symbols) and modeled (lines) normalized PDPs of the PIN photodiode CMOS SPAD in dependence on excess bias voltage for four different wavelengths. © 2021 IEEE. Reprinted, with permission, from [76].

Up to this point, we have considered light incidence perpendicular to the silicon surface. However, the incident light beam may come from another direction in many sensing and data transmission applications. The model was used to investigate the dependence of the transmission in the isolation and passivation stack and that of the PDP on the incidence angle. Figure 1.67 presents the results for a local PDP maximum at 637 nm and a local minimum at 630 nm in figure 1.63. The avalanche probability is almost independent of the incidence angle θ_0 (deviation from the silicon surface's normal), as will be reasoned below. However, the photon absorption probability changes with the incidence angle due to transmission (and reflection). As a consequence, the PDP decreases with increasing θ_0. The angle of the light beam in the silicon region θ_{Si} can be calculated by applying Snell's law at each layer interface. The bottom part of figure 1.67 plots the dependence of θ_{Si} on θ_0. The trajectory of the light in the silicon region becomes longer for increasing θ_0 and θ_{Si}, leading to a decrease of the average absorption depth. This effect is, however, quite negligible, because the refractive index of silicon is much larger than those of silicon oxide and silicon nitride, and therefore θ_{Si} is much smaller than θ_0. Even for $\theta_0 = 60°$, the trajectory changes by only 3% and the average absorption depth barely changes. Consequently, the avalanche-triggering probability can be considered to be independent of θ_0. However, due to reflection, the light intensity in the silicon decreases with increasing θ_0, and the PDP decreases accordingly (figure 1.67).

The PDP curves reflect the maxima and minima of the optical transmission of the isolation and passivation stack when the incidence angle varies. In particular, this behavior affects the bit error rate (BER) in optical wireless communications (OWC) and visible light communications (VLC) for non-perpendicular light incidence with SPAD receivers. SPAD receivers should be equipped with an antireflective coating above the SPAD to avoid a reduction of the receiver's field of view, as shown for APD receivers in [78].

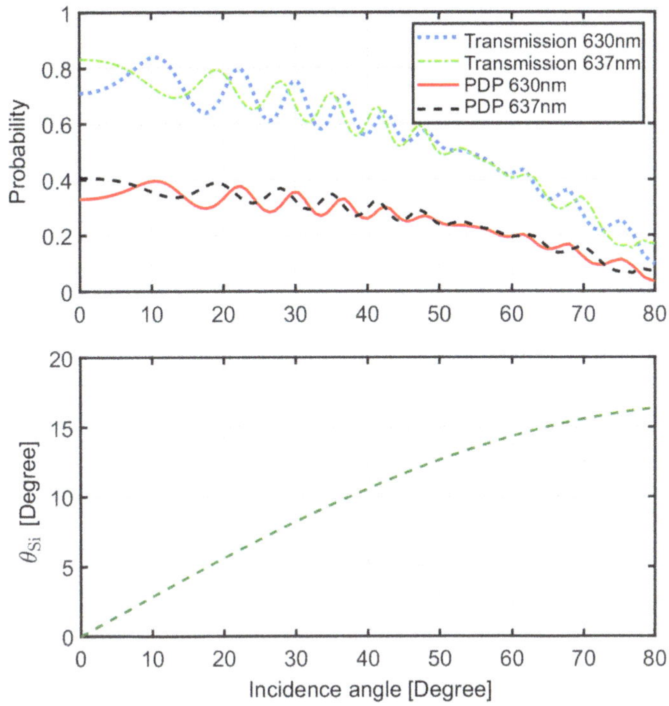

Figure 1.67. Simulated optical transmission and PDP of the PIN photodiode CMOS SPAD for two different wavelengths (top) and dependence of the angular deviation from the surface normal in the silicon θ_{Si} (bottom) on the light's incidence angle, θ_0. © 2021 IEEE. Reprinted, with permission, from [76].

The PDP model was also used to investigate the dependence of the PDP on the radial position of the light incidence and on the kind of guard ring at the cathode boundary. Two SPADs in 0.35 μm PIN photodiode CMOS (see figure 1.68) were compared in [79]. The physical n-well guard ring of SPAD1 in figure 1.68(a) had a width of 5 μm. SPAD2 in figure 1.68(b) did not have this n-well. This virtual guard ring is possible because the edge breakdown between the n+ region and the low p-doped epitaxial layer happens at a very high voltage, which occurs because the p-well diameter is smaller than the n+ diameter, i.e. the edge of the n+ region is completely surrounded by the lightly doped p− epitaxial layer. This virtual guard ring prevents edge breakdown and allows area breakdown to be exploited at the n+/p-well junction.

However, the guard ring influences the PDP in the outer part of the light-sensitive area; this conclusion was obtained using the PDP model. Two-dimensional device simulations using cylindrical coordinates were performed with ATLAS, which considered the layouts, information from the design kit, and the doping profiles made available by the ASIC foundry as confidential data. The same parameters as those listed in table 1.4 were used. The electric field obtained with the Geiger-mode feature of ATLAS for an excess bias voltage of 6.6 V is depicted in figure 1.69. Parts (a) and (b) present 2D vector plots of the electric field for the physical and virtual

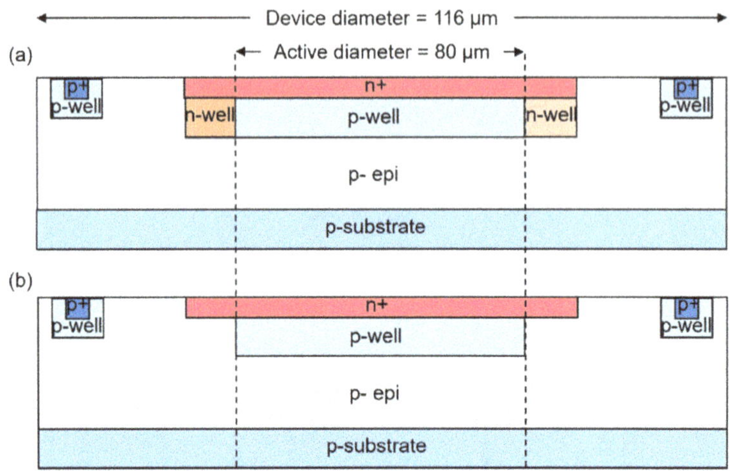

Figure 1.68. Cross sections of SPADs with (SPAD1) a physical n-well guard ring (a), and SPAD2 with a virtual guard ring (b). Reproduced from [79] with permission from SPIE.

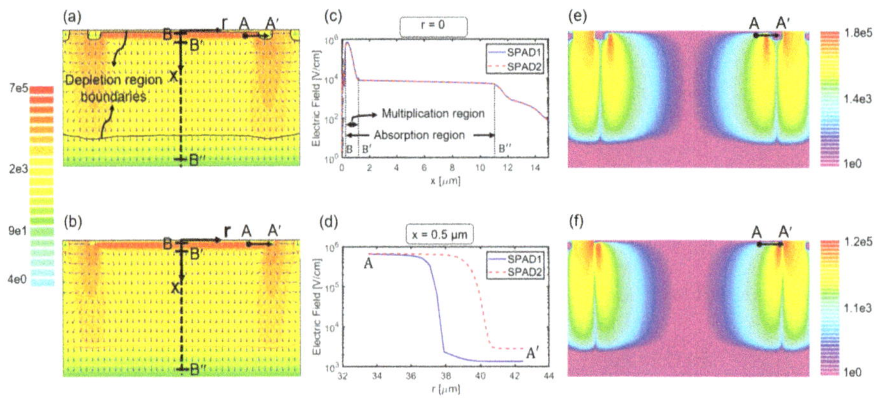

Figure 1.69. Electric field calculated using ATLAS. 2D vector plots for SPAD1 (a) and SPAD2 (b), vertical cross section at $r = 0$ (c), lateral cross section from A to A' at a depth of $0.5\,\mu$m, and logarithmic 2D plots of the radial component of the electric field for SPAD1 (e) and SPAD2 (f). Reproduced from [79] with permission from SPIE.

guard ring structures, respectively. In the center of both SPADs, the electric field distributions are similar in direction and strength. Figure 1.69(c) compares both fields in the vertical direction along B–B" ($r = 0$). When we examine the electric field inside the multiplication region at a depth of $0.5\,\mu$m between A and A' in the radial direction in figure 1.69(d), we observe a strong difference, which results from the different guard ring structures. The electric field strength is already dropping at a radial position about $2\,\mu$m smaller for the SPAD with the physical guard ring, leading to a smaller light-sensitive area. Additionally, the radial component of the electric field at the edge of the p-well increases because of the lateral p-well/n-well junction (see figure 1.69(e) and (f)). As a consequence, carriers photogenerated below A–A' are subject to a non-vertical electric field. Therefore, the electrons are accelerated into the region with a smaller electric field, where the multiplication process is weaker or absent. The active area of SPAD1 is therefore smaller than the layout area of the n+/p-well junction.

To illustrate the differences between SPAD1 and SPAD2 further, the avalanche-triggering probabilities (ATPs) obtained using ATLAS TCAD simulations are compared in figure 1.70. The ATP is the probability that a self-sustaining avalanche will be triggered by an electron or a hole that is photogenerated at a position (x, r) in cylindrical coordinates. The ATP depends on the impact ionization coefficients and the electric field. Note that the ATP is independent of the light wavelength. The PDP is determined by the ATP, as explained in the PDP model described above. Although the avalanche process only occurs in the multiplication zone, carriers photogenerated outside this zone can drift or diffuse into the multiplication zone and trigger a self-sustaining avalanche there. This is clearly visible in figure 1.70, in which the ATP region with very high values extends far below the multiplication zone through the p-well and the lightly doped epitaxial layer. Electrons photogenerated at any depth x below the multiplication zone drift upwards and have the whole thickness of the multiplication zone available for impact ionization. Holes photogenerated above the multiplication zone move downwards and also have the whole thickness of the multiplication zone available for impact ionization. For SPAD1 (top part of the figure), the ATP reaches its maximum value for carriers photogenerated inside a circle with a radius of 20.5 μm. However, SPAD2, with the virtual guard ring, achieves the same maximum ATP value inside a circle with a radius of

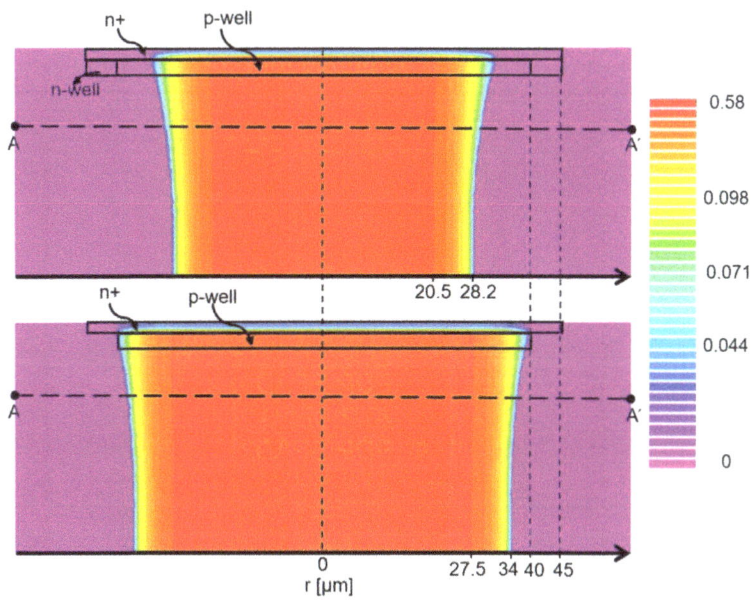

Figure 1.70. Two-dimensional plots of the total avalanche-triggering probability of SPAD1 (top) and SPAD2 (bottom); both plots are for an excess bias voltage of 6.6 V. Reproduced from [79] with permission from SPIE.

27.5 μm. The effective active area of the SPAD with the virtual guard ring is therefore 45% larger, i.e. considerably larger.

The difference between these two radii for the maximum ATP is 7 μm, which is larger than the difference between the radii of the high-field areas of 2 to 3 μm in figure 1.69(d). This is because of the lateral electric field due to the lateral p-well/n-well junction of SPAD1. This lateral electric field is stronger than that at the (lateral) p-well/p-epitaxial transition, affecting the electron's trajectory towards the cathode more strongly and narrowing the region with a high ATP in SPAD1.

For an experimental verification of the effectivity of the virtual guard ring, the radial dependence of the PDP was measured using a 635 nm light source. The measured PDPs are compared in figure 1.71 for both types of SPAD. The results verify the difference predicted by the simulations shown above. For comparison, the ATP cross sections at a depth of 5 μm are added to this figure. It should be mentioned that the slopes of the measured PDP curves are smaller than those of the ATP curves, because the measured PDP is averaged over the light spot area that was present during the measurements.

The PDP model was extended by a module that contained the parameters of an antireflective coating, and its influence on the PDP was investigated [80]. The cross section of the SPAD shown in the upper part of figure 1.47 is appropriate for this investigation. The ARC of the fabricated SPAD consisted of a 44nm thick silicon nitride layer with a refractive index $n \approx 2.0$, which is optimal for the blue spectral range. Figure 1.72 compares the spectral PDP of this HV CMOS SPAD at an excess bias of 6.6 V for devices with and without an ARC. As expected, the PDP is enhanced by the ARC for short wavelengths. This enhancement extends to about 600 nm. For wavelengths longer than 600 nm, the PDP of the device with the ARC remains below the maximum PDP of the device with the standard isolation and passivation stack, because the ARC is optimized for short wavelengths and quarter-

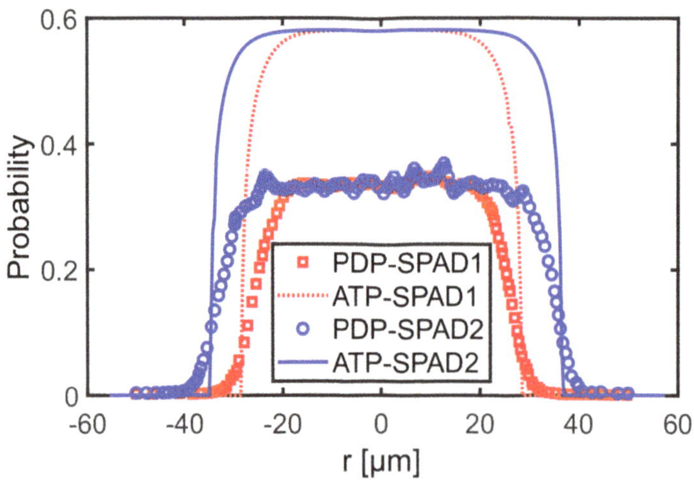

Figure 1.71. Measured radial photon detection probabilities and simulated avalanche-triggering probabilities at an excess bias voltage of 6.6 V. Reproduced from [79] with permission from SPIE.

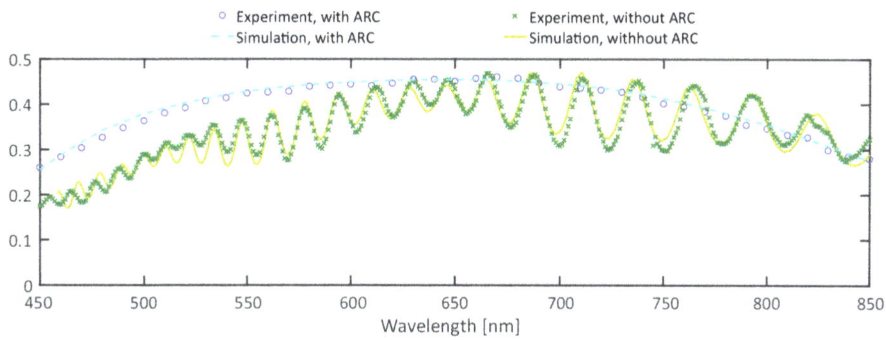

Figure 1.72. Measured and simulated PDP spectra at an excess bias voltage of 6.6 V with and without an ARC [80].

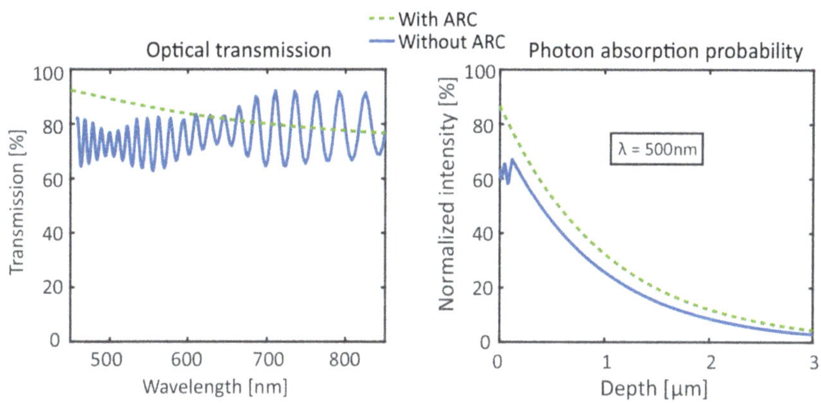

Figure 1.73. Optical transmission (left) and photon absorption probability (right) in the silicon region, both with and without an ARC at a wavelength of 500 nm [80].

wave matching is therefore not effective at these longer wavelengths. The simulated results show that the model accurately reproduces the measured results for both these cases. Therefore, we can assume that the model and the calibration of the parameters are reliable and we can investigate effects and properties that cannot be observed experimentally.

As expected, because the ARC layer is much thinner than the wavelength of the investigated spectral range, there is no standing wave inside the ARC. The left part of figure 1.73 compares the optical transmission spectra of a device with an ARC and a device with the standard isolation and passivation stack. Standing waves are present in the thick isolation and passivation stack of the reference SPAD due to the optical interference caused by multiple light reflections at the interfaces, as already seen above. The ARC layer increases the transmission for short wavelengths by reducing the reflection losses. In addition, there is no ripple in the photon absorption probability curve in the silicon for the device with the ARC (see the right part of figure 1.73) due to the absence of a standing wave; the photon absorption curve for this device follows the ideal exponential absorption profile.

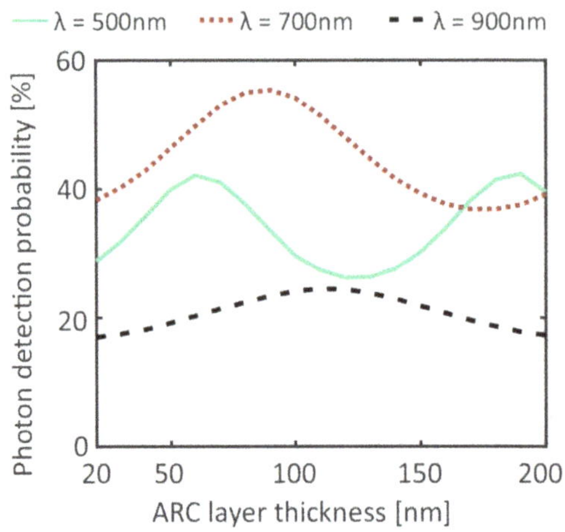

Figure 1.74. Dependence of PDP spectra on ARC layer thickness at an excess bias voltage of 6.6 V [80].

Since we observed above that the ARC can only enhance the PDP in a certain part of the spectrum, let us calculate the dependence of the PDPs for three different wavelengths on the ARC thickness. Figure 1.74 shows the PDP results for 500 nm, 700 nm, and 900 nm. There are optimum ARC thicknesses for PDP maxima at these wavelengths. These maxima occur for the optimum thickness $t_{opt}=\lambda/(4n_{ARC})$ or odd integer multiples thereof. On the other hand, PDP minima are obtained for $t_{ARC}=\lambda/(2n_{ARC})$ or integer multiples thereof. We also can see from this figure that the ARC layer thickness should be controlled to an accuracy of at least to 10% in order to keep the PDP near its maximum, if the ARC layer thickness is optimized for a certain wavelength.

In summary, an accurate PDP model was introduced, which allows the prediction of the spectral PDP and the excess bias dependence of SPAD PDPs in standard CMOS processes with the standard isolation and passivation stack and with an ARC. This model can reduce the time and costs needed for characterization and optimization in many CMOS SPAD applications. The PDP model can also be applied to non-perpendicular light incidence. Furthermore, it is astonishing that decades of use of the Lambert–Beer exponentially decaying photogeneration inside TCAD device simulators passed before a precise model for photogeneration was derived.

References

[1] Palik E D 1985 *Handbook of Optical Constants of Solids* (Orlando, FL: Academic)

[2] Aspnes D E and Studna A A 1983 *Phys. Rev.* B **27** 985–1009

[3] Sze S M 1981 *Physics of Semiconductor Devices* (New York: Wiley)

[4] Selberherr S 1984 *Analysis and Simulation of Semiconductor Devices* (Wien: Springer)

[5] Zimmermann H 2010 *Integrated Silicon Optoelectronics* (Berlin: Springer)

[6] Silvaco Int. ATLAS Manual [Online]

[7] Giudice A, Ghioni M, Biasi R, Zappa F, Cova S, Maccagnani P and Gulinatti A 2007 High-rate photon counting and picosecond timing with silicon-SPAD based compact detector modules *J. Modern Opt.* **54** 225–37

[8] 2019 http://www.micro-photon-devices.com/MPD/media/Datasheet/PDM.pdf

[9] Gulinatti A, Rech I, Panzeri F, Cammi C, Maccagnani P, Ghioni M and Cova S 2012 New silicon SPAD technology for enhanced red-sensitivity, high-resolution timing and system integration *J. Modern Opt.* **59** 1489–99

[10] https://http://www.excelitas.com/product/spcm-aqrh

[11] McIntyre R 1985 Recent developments in silicon avalanche photodiodes *Measurement* **3** 146–52

[12] Henri Dautet P, Deschamps B, Dion A D, MacGregor D, MacSween R J, McIntyre C, Trottier and Paul P 1993 Photon counting techniques with silicon avalanche photodiodes *Appl. Opt.* **32** 3894–900

[13] Stipčević M, Wang D and Ursin R 2013 Characterization of a commercially available large area, high detection efficiency single-photon avalanche diode *J. Lightwave Technol.* **31** 3591–6

[14] Ceccarelli F, Acconcia G, Gulinatti A, Ghioni M, Rech I and Osellame R 2021 Recent advances and future perspectives of single-photon avalanche diodes for quantum photonics applications *Adv. Quantum Technol.* **4** 2000102

[15] http://www.lasercomponents.com/fileadmin/user_upload/home/Datasheets/lcp/count-series.pdf

[16] Stipčević M, Christensen B G, Kwiat P G and Gauthier D J 2016 Advanced active quenching circuits for single-photon avalanche photodiodes *Proc. SPIE* **9858** 98580R

[17] Stipčević M, Christensen B, Kwiat P and Gauthier D 2017 An advanced active quenching circuit for ultra-fast quantum cryptography *Opt. Exp.* **25** 21861–76

[18] Serra N, Ferri A, Gola A, Pro T, Tarolli A, Zorzi N and Piemonte C 2013 Characterization of new FBK SiPM technology for visible light detection *J. Inst* **8** P03019

[19] Acerbi F, Paternoster G, Gola A, Zorzi N and Piemonte C 2018 Silicon photomultipliers and single-photon avalanche diodes with enhanced NIR detection efficiency at FBK *Nucl. Instrum. Methods Phys. Res.* A **912** 309–14

[20] https://hub.hamamatsu.com/jp/en/technical-note/how-sipm-works/index.html#.

[21] Marano D *et al* 2013 Improved SPICE electrical model of silicon photomultipliers *Nucl. Instrum. Methods Phys. Res. Sect.* A **726**

[22] https://http://www.hamamatsu.com/eu/en/index.html

[23] http://www.hamamatsu.com/resources/pdf/ssd/s14160-1310ps_etc_kapd1070e.pdf

[24] http://www.hamamatsu.com/resources/pdf/ssd/s14160_s14161_series_kapd1064e.pdf

[25] http://www.hamamatsu.com/resources/pdf/ssd/s13360_series_kapd1052e.pdf

[26] http://www.hamamatsu.com/resources/pdf/ssd/s13362_series_kapd1059e.pdf

[27] http://www.hamamatsu.com/resources/pdf/ssd/mppc_kapd0006e.pdf

[28] http://www.hamamatsu.com/resources/pdf/ssd/c14452_series_kacc1267e.pdf

[29] http://www.hamamatsu.com/resources/pdf/ssd/c14455-1550ga_etc_kacc1283e.pdf

[30] http://www.hamamatsu.com/resources/pdf/ssd/c13365_series_kacc1227e.pdf

[31] Villa F *et al* 2014 CMOS SPADs with up to 500 μm diameter and 55% detection efficiency at 420 nm *J. Modern Opt.* **61** 102–15

[32] Veerappan C and Charbon E 2014 A substrate isolated CMOS SPAD enabling wide spectral response and low electrical crosstalk *IEEE J. Sel. Topics Quantum Electron.* **20** 299–305

[33] Katz A, Blank T, Fenigstein A, Leitner T and Nemirovsky Y 2019 Active-reset for the n+p single-ended SPAD used in the NIR LiDAR receivers *IEEE Trans. Electron Dev.* **66** 5191–5

[34] Leitner T *et al* 2013 Measurements and simulations of low dark count rate single photon avalanche diode device in a low voltage 180-nm CMOS image sensor technology *IEEE Trans. Electron Dev.* **60** 1982–8

[35] Gramuglia F, Wu M-L, Bruschini C, Lee M-J and Charbon E 2021 A low-noise CMOS SPAD pixel with 12.1 ps SPTR and 3 ns dead time *IEEE J. Sel. Topics Quantum Electron. (Early Access)*

[36] Huang L, Wu J, Wang J, Tsai C M, Huang Y H, Wu D R and Lin S D 2017 Single-photon avalanche diodes in 0.18-μm high-voltage CMOS technology *Optics Exp.* **25** 13333–9

[37] Gersbach M, Richardson J, Mazaleyrat E, Hardillier S, Niclass C, Henderson R, Grant L and Charbon E 2009 A low-noise single-photon detector implemented in a 130 nm CMOS imaging process *Solid-State Electron.* **53** 803–8

[38] Richardson J A, Grant L A and Henderson R K 2009 Low dark count single-photon avalanche diode structure compatible with standard nanometer scale CMOS technology *IEEE Photon. Technol. Lett.* **21** 1020–2

[39] https://www.st.com/content/st_com/en/about/innovation---technology/imaging.html

[40] Pellegrini S, Rae B, Pingault A, Golanski D, Jouan S, Lapeyre C and Mamdy B 2017 Industrialised SPAD in 40 nm technology *2017 IEEE International Electron Devices Meeting (IEDM)* 16.5.1–4

[41] Zimmermann H, Steindl B, Hofbauer M and Enne R 2017 Integrated fiber optical receiver reducing the gap to the quantum limit *Sci. Rep.* **7** 2652

[42] Brandl P, Schidl S and Zimmermann H 2014 PIN photodiode optoelectronic integrated receiver used for 3-Gb/s free-space optical communication *IEEE J. Sel. Topics Quantum Electron.* **20** 6000510

[43] Steindl B, Gaberl W, Enne R, Schidl S, Schneider-Hornstein K and Zimmermann H 2014 Linear mode avalanche photodiode with 1-GHz bandwidth fabricated in 0.35 μm CMOS *IEEE Photon. Technol. Lett.* **26** 1511–514

[44] Webster E A G, Grant L A and Henderson R K 2012 A high-performance single-photon avalanche diode in 130-nm cmos imaging technology *IEEE Electron Device Lett.* **33** 1589–91

[45] Webster E A G, Richardson J A, Grant L A, Renshaw D and Henderson R K 2012 A single-photon avalanche diode in 90-nm CMOS imaging technology with 44% photon detection efficiency at 690 nm *IEEE Electron Device Lett.* **33** 694–6

[46] Mandai S, Fishburn M W, Maruyama Y and Charbon E 2012 A wide spectral range single-photon avalanche diode fabricated in an advanced 180 nm cmos technology *Optics Exp.* **20** 5849–57

[47] Niclass C, Matsubara H, Soga M, Ohta M, Ogawa M and Yamashita T 2015 A NiR-sensitivity-enhanced single-photon avalanche diode in 0.18 μm CMOS *2015 International Image Sensor Workshop (IISW)* 1–4

[48] Takai I, Matsubara H, Soga M, Ohta M, Ogawa M and Yamashita T 2016 Single-photon avalanche diode with enhanced NIR-sensitivity for automotive LiDAR systems *Sensors* **16** 459

[49] Stipcevic M, Wang D and Ursin R 2013 Characterization of a commercially available large area, high detection efficiency singel-photon avalanche diode *J. Lightwave Technol.* **31** 3591–6

[50] Rochas A *et al* 2003 Single photon detector fabricated in a complementary metal-oxide-semiconductor high-voltage technology *Rev. Sci. Instrum.* **74** 3263–70

[51] Hofbauer M, Steindl B and Zimmermann H 2018 Temperature dependence of dark count rate and afterpulsing of a singel-photon avalanche diode with an integrated active quenching circuit in 0.35 μm CMOS *J. Sensors* **2018** 1–7 9585931

[52] Goll B, Hofbauer M, Steindl B and Zimmermann H 2018 Transient response of a 0.35μm CMOS SPAD with thick absorption zone *25th IEEE Int. Conf. on Electronics Circuits and Systems (ICECS)* pp 9–12

[53] Goll B, Hofbauer M, Steindl B and Zimmermann H 2018 A fully integrated SPAD-based CMOS data-receiver with a sensitivity of −64 dBm at 20 Mb/s *IEEE Solid-State Circ. Lett.* **1** 2–5
[54] Goll B, Steindl B and Zimmermann H 2020 Avalanche transients of thick 0.35 μm CMOS single-photon avalanche diodes *Micromachines* **11** 869 (Special Issue "Miniaturized Silicon Photodetectors: New Perspectives and Applications")
[55] Steindl B, Enne R and Zimmermann H 2015 Thick detection zone single-photon avalanche diode fabricated in 0.35 μm complementary metal-oxide semiconductors *Optical Eng.* **54** 050503-1–0503-3
[56] Dervić A, Steindl B, Hofbauer M and Zimmermann H 2019 High-voltage active quenching and resetting circuit for SPADs in 0.35μm CMOS for raising the photon detection probability *Optical Eng.* **58** 40501-1–501-4
[57] Enne R, Steindl B, Hofbauer M and Zimmermann H 2018 Fast cascoded quenching circuit for decreasing afterpulsing effects in 0.35 μm CMOS *IEEE Solid-State Circ. Lett.* **1** 62–5
[58] Dervić A, Hofbauer M, Goll B and Zimmermann H 2021 Integrated fast-sensing triple-voltage SPAD quenching/resetting circuit for increasing PDP *IEEE Photonics Technol. Lett.* **33** 139–42
[59] Dervić A, Hofbauer M, Goll B and Zimmermann H 2021 High slew-rate quaduple-voltage mixed-quenching active-resetting circuit for SPAD in 0.35-μm CMOS for increasing PDP *IEEE Solid-State Circ. Lett.* **4** 18–21
[60] Steindl B, Jukic T and Zimmermann H 2017 Optimized silicon CMOS reach-through avalanche photodiode with 2.3-GHz bandwidth *Optical Eng.* **56** 110501-1–0501-3
[61] Hofbauer M, Steindl B, Schneider-Hornstein K and Zimmermann H 2020 Performance of high-voltage CMOS single-photon avalanche diodes with and without well-modulation technique *Optical Eng.* **59** 040502-1–0502-8
[62] Dautet H, Deschamps P, Dion B, MacGregor A D, MacSween D, McIntyre R J, Trottier C and Webb P P 1993 Photon counting techniques with silicon avalanche photodiodes *Appl. Opt.* **32** 3894–900
[63] Cova S, Ghioni M, Lacaita A, Samori C and Zappa F 1996 Avalanche photodiodes and quenching circuits for single-photon detection *Appl. Opt.* **35** 1956–76
[64] Burri S, Bruschini C and Charbon E 2017 LinoSPAD: a compact linear SPAD camera system with 64 FPGA-based TDC modules for versatile 50 ps resolution time-resolved imaging *Instruments* **1** 6
[65] Gulinatti A, Rech I, Panzeri F, Cammi C, Maccagnani P, Ghioni M and Cova S 2012 New silicon SPAD technology for enhanced red-sensitivity, high-resolution timing and system integration *J. Modern Opt.* **59** 1489–99
[66] Bronzi D *et al* 2012 Low-noise and large-area CMOS SPADs with timing response free from slow tails *European Solid-State Device Research Conference (ESSDERC)* 230–33
[67] Acerbi F, Cazzanelli M, Ferri A, Gola A, Pavesi L, Zorzi N and Piemonte C 2014 High detection efficiency and time resolution integrated-passive-quenched single-photon avalanche diodes *IEEE J. Sel. Topics Quantum Electron.* **20** 268–75
[68] Ceccarelli F *et al* 2018 152-dB dynamic range with a large-area custom-technology single-photon avalanche photodiode *IEEE Photonics Technol. Lett.* **30** 391–4
[69] Enne R, Steindl B and Zimmermann H 2015 Improvement of CMOS-integrated vertical APDs by applying lateral well modulation *IEEE Photonics Technol. Lett.* **27** 1907–10
[70] Bronzi D, Villa F, Tisa S, Tosi A and Zappa F 2016 SPAD figures of merit for photon-counting, photon-timing, and imaging applications: A review *IEEE Sensors J.* **16** 3–12

[71] Kohneh Poushi S S, Mahmoudi H, Steindl B, Hofbauer M and Zimmermann H 2020 Comprehensive modeling of photon detection probability in CMOS-based single-photon avalanche diodes *2020 IEEE SENSORS* 1–4

[72] Mazzillo M, Piazza A, Condorellim G, Sanfilipo D, Fallica G, Billotta S, Belluso M, Bonanno G, Cosentino L and Pappalardo A 2008 Quantum detection efficiency in Geiger mode avalanche photodiodes *IEEE Trans. Nucl. Sci.* **55** 3620–5

[73] McIntyre R J 1973 On the avalanche initiation probability of avalanche diodes above the breakdown voltage *IEEE Trans. Electron Dev.* **20** 637–41

[74] Xu Y, Xiang P, Xie X and Huang Y 2016 A new modeling and simulation method for important statistical performance prediction of single photon avalanche diode detectors *Semiconductor Sci. Technol.* **31** 065024

[75] Hsieh C-A, Tsai C-M, Tsui B-Y, Hsiao B-J and Lin S-D 2020 Photon-detection-probability simulation method for CMOS single-photon avalanche diodes *Sensors* **20** 436

[76] Mahmoudi H, Kohneh Poushi S S, Steindl B, Hofbauer M and Zimmermann H 2021 Optical and electrical characterization and modeling of photon detection probability in CMOS single-photon avalanche diodes *IEEE Sensors J.* **21** 7572–80

[77] van Overstraeten R and de Man H 1970 Measurement of the ionization rates in diffused silicon p-n junctions *Solid-State Electron.* **13** 583–608

[78] Milovančev D, Jukić T, Steindl B, Brandl P and Zimmermann H 2017 Optical wireless communication using a fully integrated 400μm diameter APD receiver *J. Eng.* **2017** 506–511

[79] Kohneh Poushi S S, Mahmoudi H, Hofbauer M, Steindl B, Schneider-Hornstein K and Zimmermann H 2021 Experimental and simulation study of fill-factor enhancement using a virtual guard ring in n+/p-well CMOS single-photon avalanche diodes *Optical Eng.* **60** 067105

[80] Kohneh Poushi S S, Mahmoudi H, Hofbauer M, Steindl B and Zimmermann H 2021 Photon detection probability enhancement using an anti-reflection coating in CMOS-based SPADs *Appl. Opt.* **60** 7815–20

Michael Hofbauer, Kerstin Schneider-Hornstein and Horst Zimmermann

Chapter 2

Photon-counting modules

In this chapter, we first explain quenching techniques and introduce the passive quenching of single-photon avalanche diodes (SPADs). We then describe advanced passive quenching, including the active resetting of SPADs. The next section addresses several different principles of active quenching; we then describe photon-counting modules (PCMs) realized as discrete circuits as well as integrated PCMs. PCMs and quenching circuits for external and integrated SPADs are addressed. An overview of commercial PCMs and the state of the art in research and development is given. Trends and possibilities for excess bias enhancement are introduced. We then summarize the properties of quenching and resetting circuits designed in submicrometer to nanometer silicon technologies.

2.1 Quenching

2.1.1 Passive quenching

Although passive quenching [1] leads to long quenching and recharging times [2], it is often an interesting possibility—especially when only a small chip area is available for the quenching circuit. Only one (high-resistance) resistor is necessary per SPAD for passive quenching. In addition, in integrated circuits, a so-called active resistor, i.e. a transistor acting as resistor, allows a high resistance value to be generated in a much smaller chip area than that required by an ohmic resistor (realised with a well or with polysilicon). Figure 2.1 shows such a passive quenching circuit, which uses a quenching transistor as the active resistor. V_B allows the resistance value to be changed.

In fact, more correctly, the metal–oxide–semiconductor field-effect transistor (MOSFET) only acts as a resistor in the initial period of passive quenching. Later, it acts as a constant current source that limits the current through the SPAD [3].

doi:10.1088/978-0-7503-2584-4ch2

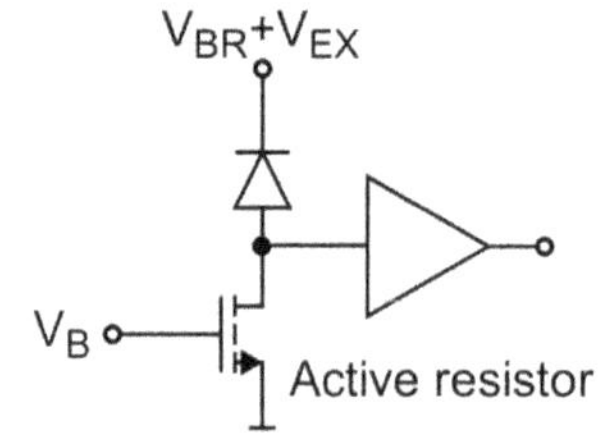

Figure 2.1. Schematic of a passive quenching circuit which uses an active resistor (a transistor used as a resistor) for quenching.

In multi-pixel SPAD sensors, in which a high optical fill factor is desirable or even a must, passive quenching is still a favourable choice. In integrated SPAD sensors with very small SPADs, for example, those described in [4–7], the SPAD capacitance is quite small and acceptable quenching and recharging times, along with dead times as low as 12ns, have been reported [8]. In a SPAD image sensor for fluorescence detection [4] the pixel pitch was 25 μm and the fill factor was 20.8% in a 0.35 μm complementary metal–oxide–semiconductor (CMOS) technology. The quarter Video Graphics Array (QVGA) SPAD image sensor reported in [5] had a pitch of 8 μm and a fill factor of 26.8%. In 0.13 μm CMOS image sensor (CIS) technology, the pixel pitch was 16 μm with a fill factor of 61% [7]. In a SPAD realised in a three-dimensionally integrated CMOS image sensor using 45nm CIS technology and pixel circuits implemented in 65 nm CMOS, the fill factor was up to 60.5% with a SPAD active area diameter of 12.5 μm [6]. A time-of-flight sensor in the same stacked technology with 128 SPADs at a pitch of 19.8 μm and with a fill factor of 31.3% was introduced in [9].

2.1.2 Advanced passive quenching

Passive SPAD quenching and recharging that uses a quenching resistor causes a long recharging time that can be reduced by an active recharge circuit. In addition, the afterpulsing probability (APP) is reduced by active recharging [10]. Figure 2.2 shows the operational principle of passive quenching with active recharging, which is also called active reset.

When the SPAD fires, a voltage drop across R_S (R_S has a low resistance) is coupled across the capacitor to the amplifier A, which triggers an output pulse to switch the N-channel MOSFET on. This MOSFET has a channel on-resistance that is much smaller than that of R_B (R_B is the passive quenching resistor), and actively charges the SPAD in a much shorter time than it is possible with passive recharging. When the MOSFET is on, the switch is opened via the inverter. This allows a defined dead time, which is set by the pulse length of the pulse generator.

For instance, multi-pixel range sensors, i.e. 3D sensors, have exploited passive quenching and active recharge (PQAR). In [11], a 128 × 128 SPAD sensor with PQAR and an integrated time-to-digital converter (TDC) in 0.35 μm high-voltage (HV) CMOS was reported. The pixel pitch was 25 μm with a fill factor of 6.16%.

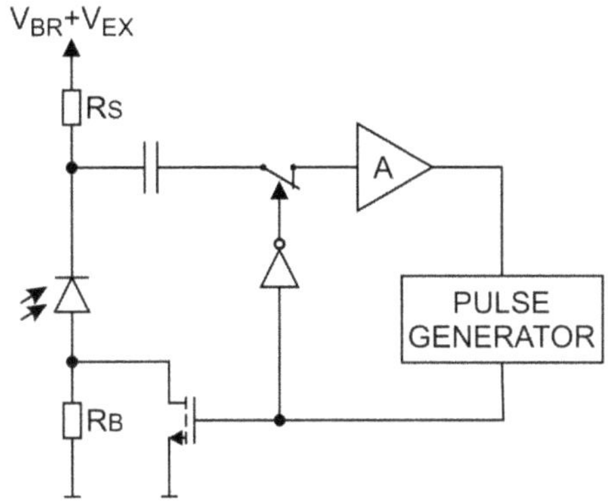

Figure 2.2. Schematic of a passive quenching circuit with active reset.

A SPAD array of 340 × 96 pixels with a SPAD pitch of 25 μm was introduced in [12]. This sensor chip was implemented in 0.18 μm HV CMOS technology. Twelve SPADs were arranged in each macro pixel with a fill factor of 70%. The SPADs used PQAR.

A 202 × 96 SPAD 3D sensor in 0.18 μm CMOS with PQAR was presented in [13]. The optical fill factor was again, 70%. The SPADs were organized into 6 × 4 subarrays.

A 64 × 64 direct time-of-flight (TOF) SPAD sensor with very impressive performance was reported in a so-called digital silicon photomultiplier configuration [14]. A 0.15 μm CMOS process was used. Each pixel, which had a dimension of 60 μm, contained eight SPADs and the fill factor was 26.5%. A quenching MOSFET (active resistor) and a clamping transistor with thicker gate oxide supported an excess bias of 3.3 V and a connection to the low-voltage digital input with 1.8 V transistors.

2.1.3 Active quenching

For active quenching, the avalanche event has to be detected very rapidly by a transistor, an inverter, or a comparator. This detecting device then has to switch on a quenching transistor with a very short delay. This quenching transistor discharges the SPAD to its breakdown voltage or less. Figure 2.3 shows the active quenching principle.

Active quenching can be faster than passive quenching. In fact, with active quenching circuits, in the first phase after the absorption of a photon and the start of an avalanche buildup, passive quenching occurs until the detection threshold of the circuit is reached—actually, even longer, by the length of the delay or reaction time of the active quenching circuit, until its output really pulls the SPAD below its breakdown voltage. Therefore, strictly speaking, active quenching circuits use mixed quenching. To make active quenchers faster than passive quenching, so-called active reset is used to rapidly charge the SPAD to a voltage above the breakdown voltage again. The recharging switch shown in figure 2.3 fulfills this task.

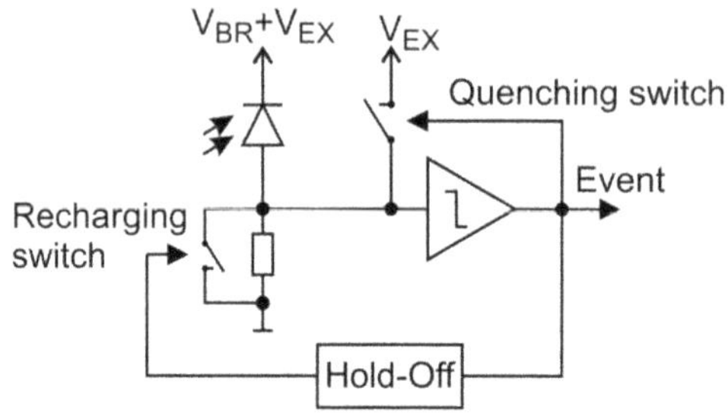

Figure 2.3. Schematic of an active quenching circuit with active reset.

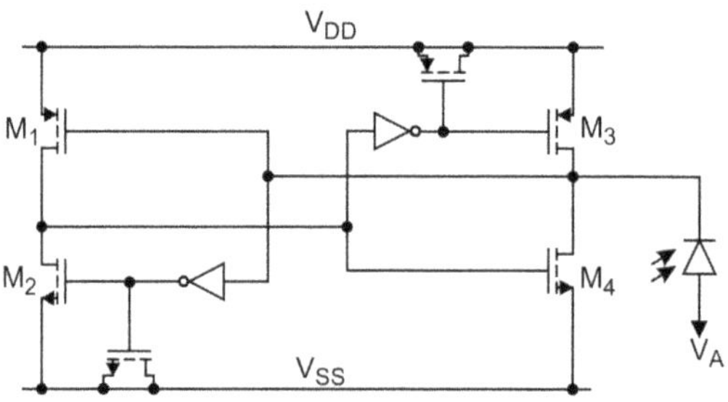

Figure 2.4. Schematic of a dynamic, thyristor-like quenching circuit.

Active quenching was, for instance, applied in a SPAD array for Raman spectroscopy [15]. The technology was 0.35 μm HV CMOS, and thin SPADs were implemented.

A regenerative latch principle was suggested for active quenching [16]. Figure 2.4 shows the so-called dynamic, thyristor-based quenching circuit. In the waiting state, the SPAD is biased close to V_{DD}, because M2 has a larger gate width than M1, and M3's width is larger than that of M4. Therefore, the leakage currents keep the SPAD biased close to V_{DD}. When an avalanche occurs, passive quenching starts via M3. However, M1 and in turn M4 begin to conduct when the voltage drop across M3 exceeds the threshold voltage of M1. M4 pulls the cathode down to V_{SS} and the SPAD is actively quenched. Due to the regenerative feedback, this action is fast. In a 0.35μm CMOS, the pixel circuit occupied an active area of 130 μm^2. According to simulations, the fastest reported quenching time was 2 ns and the circuit only consumed 60μW at a 25 MHz triggering rate [16].

Another fast quenching circuit was introduced in [2]. The two main ideas behind this approach are illustrated in figure 2.5. First, the circuit senses the avalanche current, and second, the circuit breaks the current path, leading to a fast decrease of the voltage across the SPAD and therefore to fast quenching. This principle benefits from the well-known fact that current-mode circuits are faster than voltage-mode circuits [17].

A schematic of the current-mode quencher is depicted in figure 2.6. In the waiting phase, node A is close to ground. M2 is on and the output is at an HV. After

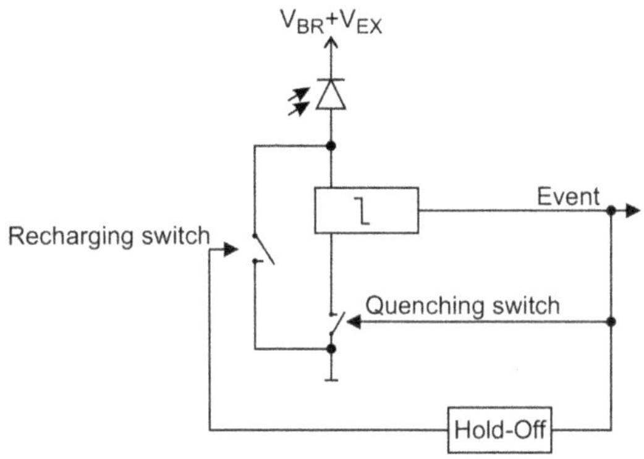

Figure 2.5. Principle of current-mode active quenching and active recharging [2].

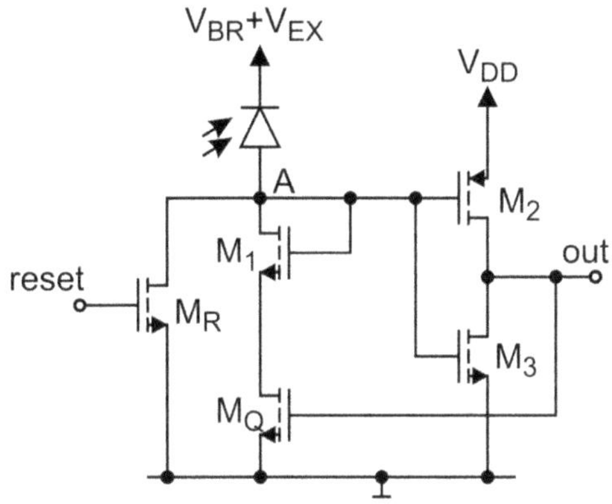

Figure 2.6. Schematics of a current-mode active quenching and active recharging circuit [2].

detection of a photon, the avalanche current flows through the MOS diode M1 and the switch MQ. MQ has low resistance, because its gate is at V_{DD} via M2. Therefore, M1 and M3 form a current mirror and the avalanche current is mirrored in the output stage. The output voltage drops. In turn, the channel resistance of MQ increases, and therefore the avalanche current leads to a fast increase of the voltage drop across M1 and MQ, which means that the voltage across the SPAD decreases rapidly, which quenches the SPAD rapidly. The positive feedback via M1, M3, and MQ speeds up this process. M3 turns on completely, which turns off MQ, and the current path for the avalanche current is broken. The SPAD then quenches itself because the avalanche current discharges its capacitance. After a hold-off time, MR is switched on, recharging the SPAD actively.

2.2 PCMs using discrete circuits

The product portfolio of Hamamatsu Photonics K.K. contains a photon-counting module [18]. The C11202-050 [19] is a discrete single-pixel photon counter (SPPC) built according to the basic block diagram shown in figure 2.7.

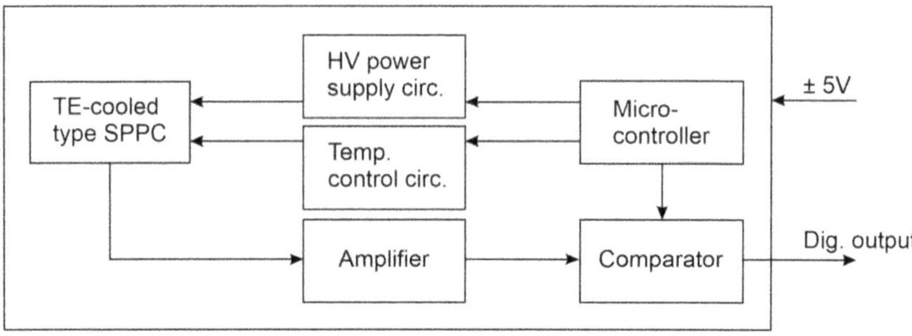

Figure 2.7. Block diagram of Hamamatsu counting modules [20].

This photosensitive device is available with active area diameters of 50 and 100 μm. The peak PDP is 70% for 450 nm light and the element temperature given is −20 °C; this leads to a low dark count rate of a maximum of 25 cps for the smaller diode and a maximum of 100 cps for the larger diode. The APP is given as 0.1% for dead times from 100 to 500 ns. The maximum count rates are 30 and 20 Mcps for the smaller and larger diodes, respectively.

Excelitas Technologies [21] also offers a series of single-photon-counting modules (SPCMs) [22]. The SPCM-AQRH series [23] offers several options in terms of dark counts, from 25 to 1500 cps at room temperature, dead times from 22 to 40 ns, a maximum count rate of 37 Mcps and a maximum APP of 1%. The maximum PDP is 70% at 700 nm. The circular diode has a diameter of 180 μm.

A dead time of 77 ns and therefore a maximum count rate of 13 Mcps are offered by the PDM series [24] of Micro Photon Devices S.r.l. (MPD) [25]. Diodes are available with active area diameters from 20 up to 200 μm. The peak PDP is 49% for 550 nm light; the DCR and APP depend on the diode diameter and vary from 5 cps for the smallest device up to less than 1000 cps for the largest device with thermoelectric cooler (TEC)-cooled SPADs. A fast gated SPAD module [26] is available, also manufactured by MPD. Its gate repetition frequency is up to 80MHz and its maximum PDP is 50% at 400 nm for an excess bias voltage of 5 V. The active area of the SPAD has a diameter of 50 μm and the DCR is given as 200 cps at 25 °C. Nevertheless, the SPAD can be cooled down to −10 °C which leads to smaller dark counts. Figure 2.8 shows the principle of the gated mode. The gate sync biases the SPAD in Geiger mode; when a photon hits, the avalanche is quenched in 1 ns and an output pulse is generated. During the adjustable hold-off time, the SPAD is kept turned off. After that, the detector is turned on according to the gate sync. Not only is the gated mode possible, but also a free-running mode, in which the SPAD is always active, except for the hold-off time after a count.

Laser Components GmbH [27] also markets a PCM series. Again, the dark count rate is selectable from 10 to 250 cps, the peak PDP is 70% at 670 nm, and the APP is 1% for a SPAD with an active diameter of 100 μm. Unfortunately, no maximum count rate is given. Different types can be chosen for different wavelengths.

The SPCM manufactured by Thorlabs Inc. [28] offers several operating modes, including not only free-running timed counting, but also an external gating mode.

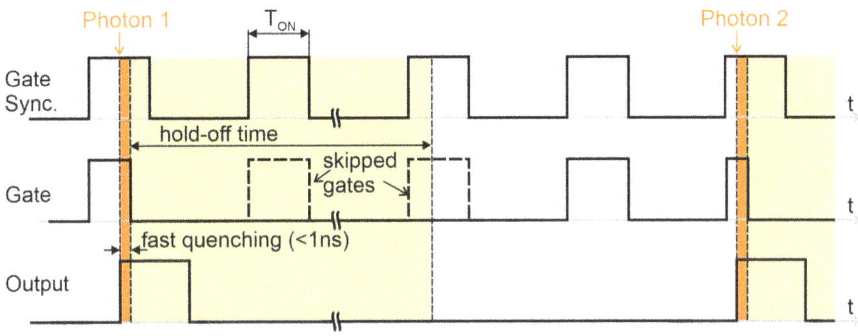

Figure 2.8. Principle of the gated mode [26].

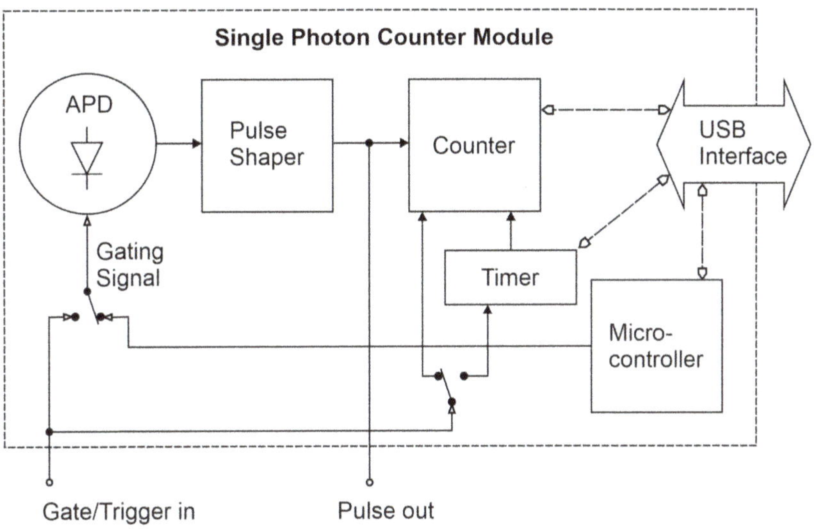

Figure 2.9. Block diagram of Thorlabs' SPCM, which supports gating mode [28].

It is possible to control the SPAD's bias voltage by an external gating voltage, so that it is not always biased in the Geiger mode. The block diagram is shown in figure 2.9. The silicon SPADs are available with 20 μm (SPCM20A) and 50 μm (SPCM50A) diameters, with DCRs of 60 cps and 200 cps, respectively. The maximum count rates are 28 Mcps for the smaller diode and 22 Mcps for the larger one; these values were reached in the externally gated mode. The peak PDP is given as 35% for 520 nm light. The APP is given as 3%.

Stanford Research Systems [29] developed a photon counter with two channels (the SR400 [30]). Count rates of up to 200 Mcps are possible with each independent channel. The inputs are internally terminated by 50 Ω, and the instrument is delivered without light-sensitive devices. Five-nanosecond pulse pairs can be resolved (table 2.1).

Table 2.1. Overview of PCMs using discrete circuits.

Ref.	Max. count rate (Mcps)	Max. PDP (%)	λ (nm)	Max. APP (%)	Max. DCR (cps)	Active diameter (μm)	Cooled (y/n)
[19]	30	70	450	0.1	25	50	y
[19]	20	70	450	0.1	100	100	y
[22]	37	70	700	1	25	180	y
[24]	13	49	550	3	5	20	y
[24]	13	49	550	3	25	50	y
[24]	13	49	550	3	1000	200	y
[26]	80	50	400		200	50	y
[27]	1	70	670	1	10	100	
[28]	28	35	500	3	5	20	y
[28]	22	35	500	3	25	50	y

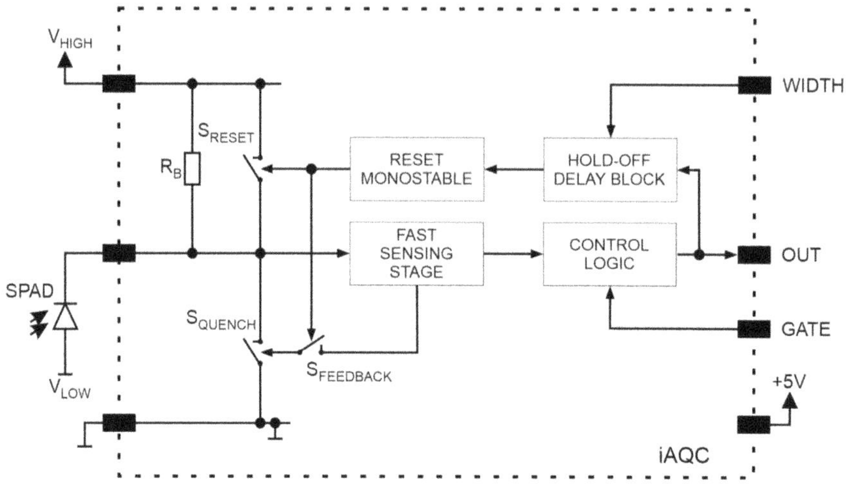

Figure 2.10. Block diagram of an active quenching circuit (AQC) [31].

2.3 PCMs using integrated circuits

In this section, we describe PCMs with integrated quenching circuits. They either have an integrated SPAD or an external SPAD.

A monolithic active quenching and reset circuit is described in [31]. The circuit is realized in an HV 0.8 μm CMOS technology with the goal of keeping the avalanche charge as low as possible for each event in order to minimize the APP, self-heating effects, and optical crosstalk in the case of detector arrays.

In idle mode, the resistor R_B in figure 2.10 pulls the cathode of the external SPAD to V_{high}, which biases the detector at a voltage greater than the breakdown voltage

V_{BD}; the voltage | V_{low} | is set to slightly less than V_{BD}. When a photon hits, the avalanche current starts flowing through R_B and lowers the input node potential, which is detected by the sensing circuit; active quenching is triggered and the SPAD is fast quenched by S_{quench}. The control logic processes the pulse and recharges the SPAD after an adjustable hold-off time. The super-low k (SLIK) diode [32] used for the experiments exhibits a V_{BD} of 439V, a V_{low} of – 435 V and a V_{high} of 20 V; this leads to an excess bias voltage of 16 V. The reported circuit therefore provides a quenching voltage of 20 V, which quenches the diode down to about 4 V below V_{BD}. One quenching cycle takes 50 ns for the minimum hold-off time and therefore results in a maximum count rate of 20 Mcps.

A SPAD was integrated together with a quenching circuit using standard 0.8 μm CMOS technology [33]. The integrated detector had a diameter of 12 μm and consisted of a p+ n-well junction with deep p-diffusion at the end of the p+ layer to avoid edge breakdown. This led to a V_{BD} of 16V, which could be applied via the p+ anode of the device, – V_{low}, see figure 2.11. The cathode voltage was sensed by a source follower, N_S. When the SPAD fired, the gate of the p-channel MOSFET P_S became more negative, P_S switched on, a positive gate–source voltage was applied to S_{quench} via $S_{feedback}$ and the SPAD was quenched.

The n-well cathode of the SPAD is pre-biased via the resistor R_B to V_{high}, which equals 10 V in this circuit. The passive–active quenching works in a similar way to the work of the same group described above. For this integrated SPAD, the minimum dead time of 30 ns leads to a count rate of 30 Mcps. The maximum detection efficiency is 40% at a wavelength of 500 nm for an overvoltage of 10 V. The DCR for this 10 V overvoltage is 35 kcps and drops significantly to 600 cps for an overvoltage of 5 V. The APP depends, of course, on the dead time; for a 55 ns

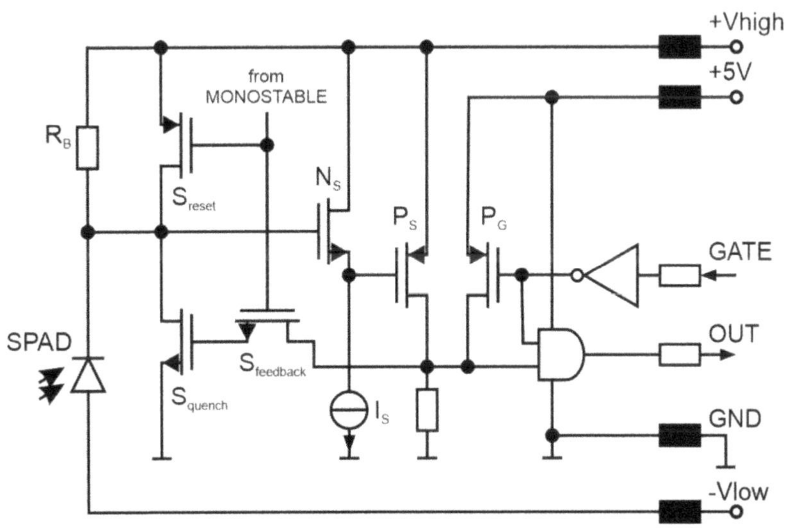

Figure 2.11. Schematic of AQC with integrated SPAD [33].

hold-off time, the APP is 2.6% at a 5 V overvoltage, and it drops to 0.02% for a 200 ns hold-off time.

Reference [34] also presents an integrated SPAD using conventional 0.8 μm CMOS technology. The circular detector has an active area of 30 μm^2 and is equipped with a mixed-mode quenching circuit. The avalanche current is passively quenched via R_{quench} (see figure 2.12), and the SPAD is actively recharged using a multivibrator circuit to recharge it back to V_{DD} via M_{res}. This multivibrator consists of a Schmitt trigger followed by an inverter with a feedback loop that uses R and C to define the oscillation frequency. When a photon arrives, the comparator switches the output of the first inverter (IV_1) to GND and turns M_{mv} off, and the multivibrator starts to oscillate. The delay time of 5ns before the recharge transistor opens ensures that the avalanche current is quenched. Recharging the SPAD to an overvoltage of 5 V again turns on the transistor M_{mv} again, and subsequently, M_{res} is opened again.

The voltage across the SPAD is defined by an external bias voltage V_{op}, which is negative in order to generate a reverse bias voltage. All the measurements presented in this work were measured with a V_{ex} of 2.5 V. The circuit can detect photons with a dead time of 10 ns and therefore it can achieve a count rate of 100 Mcps. The maximum sensitivity of the circuit is 21% at 440 nm, which drops to approximately 5% at 700 nm. The SPAD has a thin depletion region, and therefore, longer wavelengths with higher penetration depths are less likely to be absorbed. The relatively small active volume offers the advantage of a low DCR of 60 Hz at room temperature.

Another design that used 0.35 μm CMOS technology was published in [35]. The integrated photodiode described, consisting of a p+/HV n-well junction with n enrichment below the p+, described in [36], was reverse biased by an external voltage V_B (see figure 2.13). The anode of the SPAD was connected to the circuit.

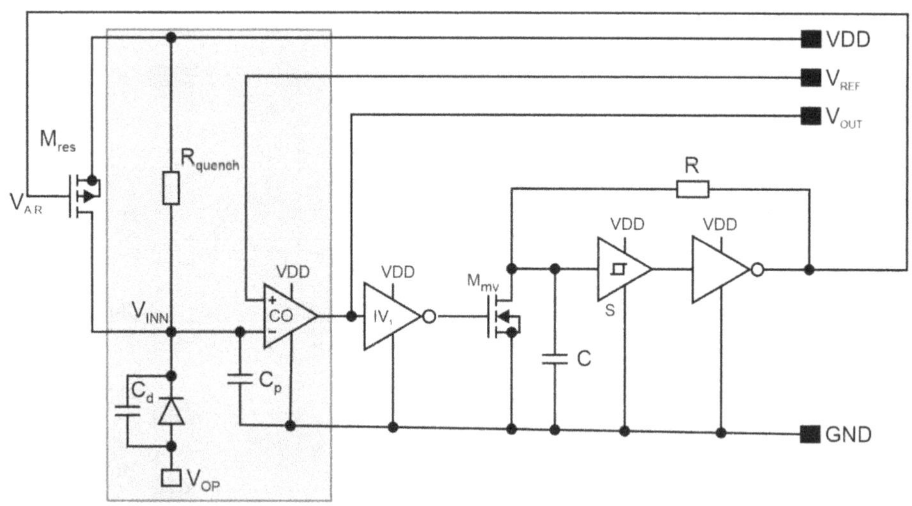

Figure 2.12. Schematic of an AQC with an integrated thin SPAD [34].

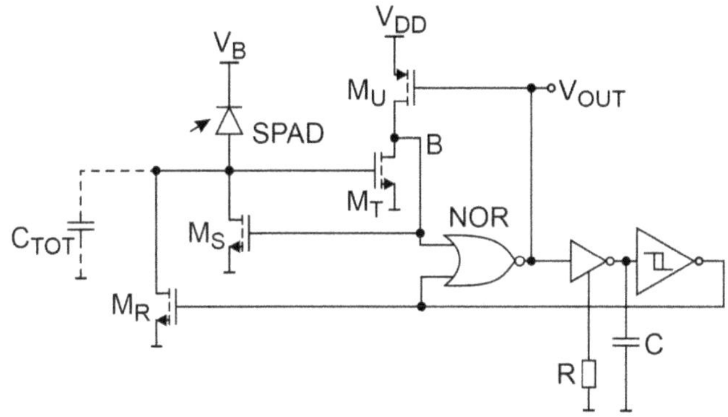

Figure 2.13. Front end with integrated SPAD; the anode is connected to the circuit [35].

The transistor M_S sensed the incoming avalanche and activated M_T, leading to a voltage drop at node B. M_S closed, and the avalanche was quenched in the process, as the current path toward ground was cut off. The subsequent logic path reset the anode voltage of the SPAD to a voltage higher than the breakdown voltage by pulling it towards ground again via M_R. The R–C combination in the logic loop defined the hold-off time; in this work, it was reported to be 20 ns, to avoid high afterpulsing, which would have been expected if the hold-off time had been shorter.

Since the described circuit was able to detect incoming photons, even in the reset phase when M_R was turned on, the maximum count rate rose to 50 Mcps. A line of sensors was described in [35]; each SPAD was connected to the described circuit. Unfortunately, the number of SPADs was not reported; all measurement data described a single pixel. The breakdown voltage was 25V and the excess bias voltage was reported to be 6 V. The round SPAD had a diameter of 20 μm. The measurements were performed with a 570 nm LED; a PDP of 28% and a DRC of 25 cps were reported at room temperature [36]. The afterpulsing probability was rather low: 1.3% for a hold-off time of 20 ns, which decreased for increasing hold-off times.

An active quench and reset circuit with a digitally adjustable hold-off time was presented in [37]. It was fabricated using 0.35μm CMOS technology by ams AG (Premstaetten, Austria) and offered the possibility of controlling the hold-off time via an eight-bit code in 6.5 ns steps from 28.4 ns to microseconds. Figure 2.14 shows the block diagram of the proposed circuit. When a photon hits, there is a voltage drop across the resistor R_S, which is sensed by the comparator (comp), V_{Cout} switches from low to high and activates the ring oscillator, also resetting the counter. Subsequently, the voltage at node Q is also set to high, opening the NMOS transistor, which quenches the SPAD rapidly towards ground. In the meantime, the oscillator pulses are counted by the eight-bit synchronous binary counter and compared to the eight-bit input word provided by the user via XNOR gates. When the counter equals the chosen value, the outputs of the XNORs go low and therefore the PMOS transistor is opened to reset the cathode of the SPAD to V_{DD} again, changing the voltage at node Q to low and closes the NMOS transistor.

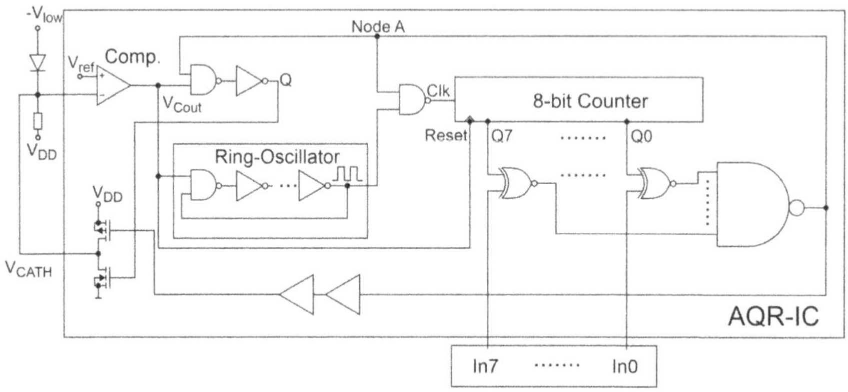

Figure 2.14. Block diagram of a digitally adjustable AQC [37].

The disadvantage of this digital approach for adjusting the hold-off time is the fact that the area consumption is rather high. The authors of [37] stated that 25% of the core circuit was used for the 41-stage ring oscillator. The excess bias voltage was given as 3 V and the maximum count rate was 35.2 Mcps.

The quenching circuit described above [37] was used in several further works, e.g. in [38], where it was used to build a microcontroller-based programmable system with a 20μm SPAD described in [39] using a 1.5 μm CMOS-compatible process. For an excess bias voltage of 1 V, the measured DCR was 80 cps and it increased to about 1 kcps for an excess bias of 3 V. The breakdown voltage was reported to be 24.6 V at room temperature. A peak PDP of 40% was achieved with 600 nm light, while the APP was 1% for a hold-off time of 100 ns, which decreased for longer dead times.

In [40], two APDs were integrated on the same chip, based on the p–n junction SPAD using the same technology described above [39]. One of the APDs was connected to the quenching circuit of [37] and driven in the Geiger mode; the second APD was used as an APD in the linear mode and connected to a two-stage transimpedance amplifier (see figure 2.15). The careful combination of these two APDs in two operating modes allowed a high dynamic range of 164.2 or a 132 dB linear dynamic range.

Reference [41] basically used the same circuit as that described above [35], but it was combined with much larger SPADs with active diameters of up to 500 μm in a 0.35 μm HV CMOS technology (see 1.3.1). Different hold-off times were reported for different SPAD diameters; they rose from 40 ns for a 20 μm diameter to 100 ns for 100 μm and even to 150 ns for a 500 μm diameter. Therefore, the maximum count rates were calculated to be 25 Mcps for 20 μm, 10 Mcps for 100 μm, and 6.7 Mcps for the 500 μm diameter.

An array of 32 channels for time-correlated single-photon-counting measurements was presented in [42]. The photosensitive part of the system was completely built in a controlled atmosphere chamber to ensure optimum SPAD behaviour. The anode of the external SPAD was bonded to an active quenching circuit, while the

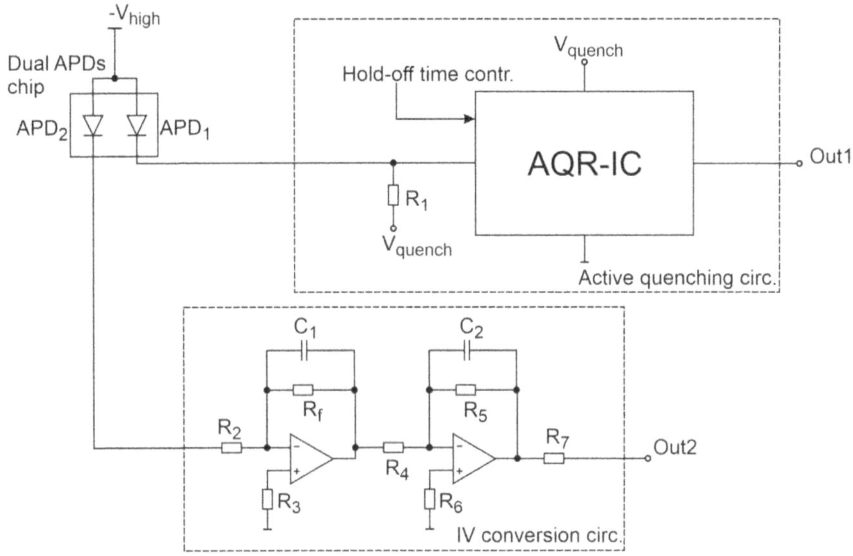

Figure 2.15. Schematic of the dual-APD circuit described in [40].

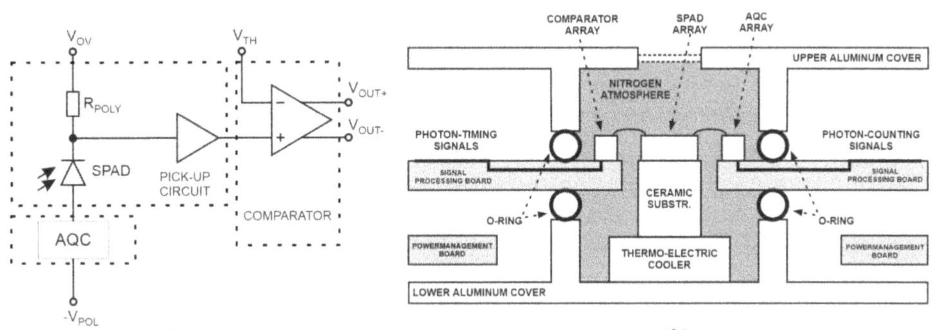

(a) (b)

Figure 2.16. (a) Block diagram of a single pixel and (b) a sketch of the controlled atmosphere chamber architecture [42].

avalanche pickup circuit was connected to the cathode of the SPAD (see figure 2.16(a)). The reason for placing the pickup and quenching circuits at different nodes of the SPAD was to avoid crosstalk between pixels, as described in [42]. The voltage drop across the SPAD was defined by V_{OV} and $-V_{POL}$. In the idle state, the SPAD was biased in the Geiger mode ($V_{SPAD} = V_{OV} + V_{POL}$); as soon as a photon triggered an avalanche, the SPAD was passively quenched by the resistor R_{POLY}. Subsequently, the integrated pickup circuit, which behaved as an inverter, activated the comparator and subsequently the anode voltage of the SPAD was increased to quench the avalanche completely. The single-pixel architecture of [42] used the same structure as [43], but the number of pixels was increased. In [43], the AQC was described in more detail; it was fabricated using a 0.35 μm CMOS technology, the same as the comparator of the pickup circuit, but on separate dies, due to the resulting easier handling of the different voltage regimes. The supply of the AQC could be regulated between 7.5 V and 18 V. When an avalanche was detected, the anode voltage of the SPAD was pulled to the supply voltage of the AQC via a pMOS transistor. Unfortunately, none of the cited papers described the circuit of the AQC in detail. The new version in [42] was built using 0.18 μm HV CMOS technology. The minimum dead time was 16 ns, which led to a maximum count rate of 60 Mcps per AQC. The SPAD bias voltage could be varied between 20 and 43 V. The excess bias voltage reached was 6 V. All three components were located in a nitrogen atmosphere, as depicted in figure 2.16(b). The top of the aluminum chamber was glass, so that photons could enter the chamber. The temperature of the SPAD could be cooled to −15 °C by a TEC in order to reduce the DCR and to keep the SPAD at a constant temperature during the measurements. The nitrogen atmosphere was necessary to avoid the formation of fog on the devices due to the cooling. The SPADs themselves and the pickup circuit were built using a custom technology. A high PDP of 44% for 550 nm light and a low DCR of less than 400 cps were achieved at −10 °C. In particular, the DCR basically improved with lower temperatures; it was reported to be 20 kcps at 25 °C for the same device. The maximum count rate of the complete system was reported to be 1 Mcount/s per channel with a SPAD diameter of 50 μm.

In [44], the same group published an 8 × 8 array of 50 μm-diameter SPADs, which were again built in a dedicated silicon technology for sensor applications. Two operational modes were proposed: first, a multi-spot configuration to detect photons at different points in the sample, and second, operation with a suitable microlens optic to focus the light on the photosensitive spots only and therefore avoid the loss of photons in the dead spaces. The AQC used in this work was the same as described above [42], but two of them were implemented, since 64 channels were needed and the AQC only provided 32 channels. The complete photon detection head was again cooled inside a nitrogen chamber similar to the version depicted in figure 2.16(b). The best single pixel of this work exhibited a performance of 33 Mcps at an excess bias voltage of 6 V; it had a low DCR of 72 cps at −10 °C, with a peak PDP of 49% for 550 nm light. Working in a combined mode, the 8 × 8 array of the complete system yielded a maximum count rate of 2130 Mcps.

Reference [45] presented an approach for increasing the PDP of SPAD sensors by combining commercial SPADs (part numbers SPCM 80024 FC 28234 and SPCM 80024 FC 38470 [46]) with an active quenching and active resetting sensor (AQAR) designed using a 0.35 μm HV bipolar CMOS–double-diffused metal–oxide semiconductor (DMOS) (BCD) technology, as shown in figure 2.17. The technology used offered the opportunity to use HV lateral double-diffused MOSFETs (LDMOS) transistors with breakdown voltages of about 100 V and diodes with breakdown voltages of about 80 V and enabled a maximum quenching voltage of 68 V. The anode of the external SPAD was connected to the IN pad and biased at a voltage greater than the breakdown voltage of the SPAD. The receiver was active when the quenching input was high, and therefore M_2 was turned on and subsequently M_1

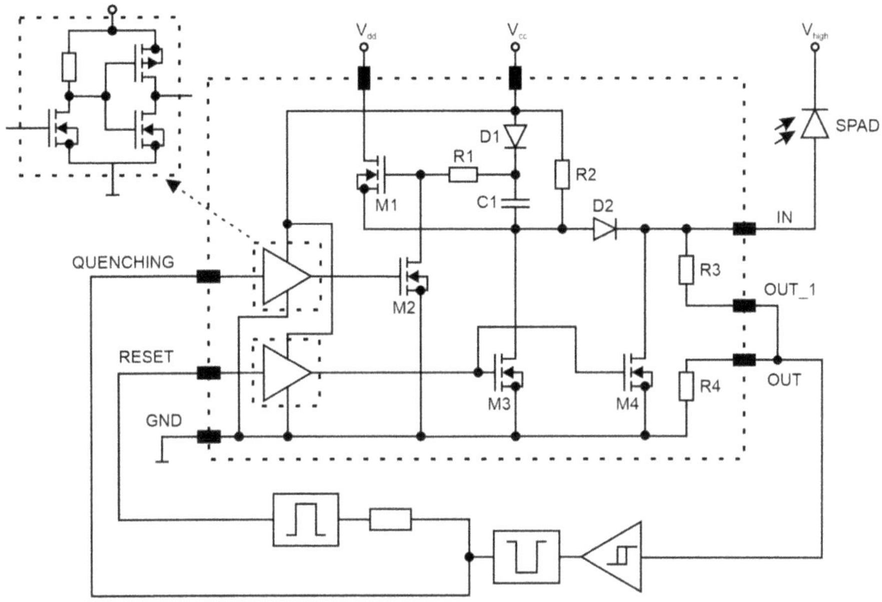

Figure 2.17. Schematic of the AQAR chip; the anode of the external SPAD is connected to the input pad [45].

was turned off. M_3 and M_4 were the reset transistors and they were turned off as well. M_1 to M_4 were LDMOS transistors that handled the high quenching voltage. When a incoming photon arrived, a voltage drop occurred across R_3 and R_4, and the output signal triggered the quenching input to become low; therefore, M_1 was activated via M_2 and actively quenched the avalanche by charging the IN node toward V_{DD}; as a result, the voltage across the SPAD was lowered to less than V_{br}. After the delay time, a reset was performed via M_3 and M_4.

Apart from the LDMOS transistors, the logic was at the transistor–transistor logic (TTL) level; the quenching and reset input voltages were boosted to an amplitude of 12 V to drive the HV transistors. The maximum quenching time measured was 30 ns for a 68 V quenching voltage and the maximum reset time was 10 ns. A minimum dead time of 42 ns was measured for the minimum hold-off time. For the measurements and a comparison with the commercial devices mentioned above, the hold-off time was set to 40 ns, which led to a dead time of 82 ns. The breakdown voltages of the SPADs used were 339.5 and 355 V, respectively, and the new quenching circuit offered a higher excess bias voltage of 68 V instead of the 30 V of the original products. This led to increased PDPs of 73.7% and 75.1% instead of 68.3% and 69.5% at 785 nm, respectively. This increase was obtained at the cost of a higher APP, which increased by roughly a factor of 2.7, and a higher DCR, which nearly doubled due to the higher excess bias voltage. The maximum count rate decreased from 31 Mcps down to 10.1 Mcps.

An integrated SPAD described above (1.3.1) in a 0.18 μm CMOS image sensor technology was combined with a passive quenching and active resetting circuit [47]. The proposed active-reset quenching circuit can be seen in figure 2.18. The

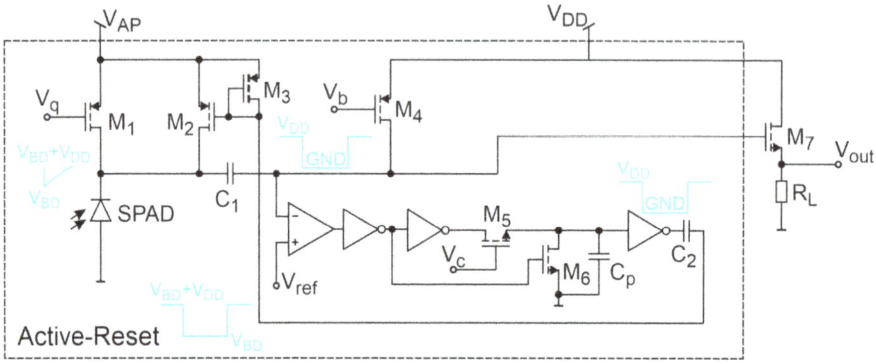

Figure 2.18. Proposed passive quenching, active-reset circuit [47].

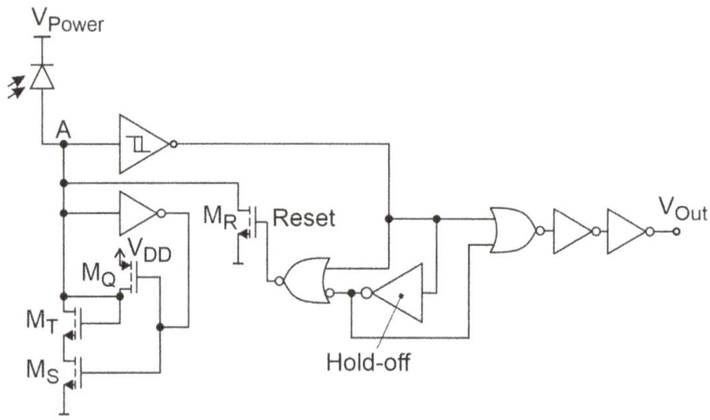

Figure 2.19. Schematic of the active quenching and recharging circuit described in [48].

quenching is performed passively via M_1. In this circuit, M_1, M_2, and M_3 form an extra voltage domain inside an n-well connected to $V_{AP} = V_{BR} + V_{EX}$. This part is separated from the rest of the circuit, which is in the voltage range from V_{DD} to ground, by the capacitances C_1 and C_2. When a photon arrives, the avalanche is passively quenched by M_1, whose 'on' resistance can be adjusted by the gate voltage V_q. The voltage pulse at the input node is detected by a fast comparator and generates a pulse on the gate of M_2 via an adjustable delay element that actively resets the SPAD. All the transistors except for M_3 and the comparator circuit are 1.8 V transistors, therefore the excess bias voltage for the active-reset circuit appears to be 1.8 V; unfortunately, this is not mentioned in [47]. Nevertheless, the maximum count rate was given as 250 Mcps and the DCR for the low excess bias voltage was less than 70 Hz.

Reference [48] presented an active quenching and recharging circuit. The circuit was designed using standard 0.18 μm CMOS technology. Figure 2.19 shows the schematic of the circuit. In the idle state, the anode of the SPAD is close to ground due to the open transistors M_S and M_T. The output of the first inverter INV_1 is therefore high and M_Q is off. The V_{out} of the Schmitt trigger is high. As soon as an avalanche occurs, the avalanche current causes a voltage drop across M_T, which is diode connected, and M_S; this flips the first inverter INV_1 which closes transistor M_S and opens M_Q. The open M_Q rapidly pulls the anode node toward V_{DD}. The Schmitt trigger changes its output as soon as the avalanche current is completely quenched and starts the recharge process after a hold-off time. M_R opens and recharges the SPAD.

The integrated SPAD consists of a p+/p-well anode and a deep n-well cathode. The active diameter of the octagonal device is 10 μm, the breakdown voltage reported was 15.5 V, and the applied excess bias voltage was 3.5 V. A peak PDP of 34% was reported at 450 nm. The corresponding DCR was about 6.9 kHz and the APP was reported to be 0.75% for a hold-off time of 4 ns. The hold-off time together with the quenching time of about 0.7 ns led to a maximum count rate of 200 Mcps.

Figure 2.20 shows the schematic of an integrated quenching circuit which is connected to the anode of an external SPAD [49]. The HV MOSFETs M_Q and M_R regulate the anode voltage of the SPAD. They are controlled by the high-side logic (HSL) and the low-side logic (LSL) for M_Q and M_R, respectively. These transistors define the maximum quenching voltage of the circuit, which is up to 50V due to the 0.18 μm HV CMOS technology used. The complete logic circuits are designed using 1.8 V transistors to benefit from the speed advantages of low-voltage transistors. An incoming photon and the subsequent avalanche are detected by the sense circuit, which consists of a low-voltage sense circuit as well as a HV switch transistor, which is connected to the anode of the SPAD and keeps the sense circuit in the safe operational range. The current through the SPAD increases the drain–source voltage of the sense transistor M_{sense} and triggers active quenching, which leads to a quenching time of 60 ns and a reset time of 40 ns for an excess bias voltage of 30 V. The maximum possible excess bias voltage is 50 V. The logic circuits are operated in their own voltage regime from V_{DD} to V_{DD} − 1.8 V for the HSL and 1.8 V to GND

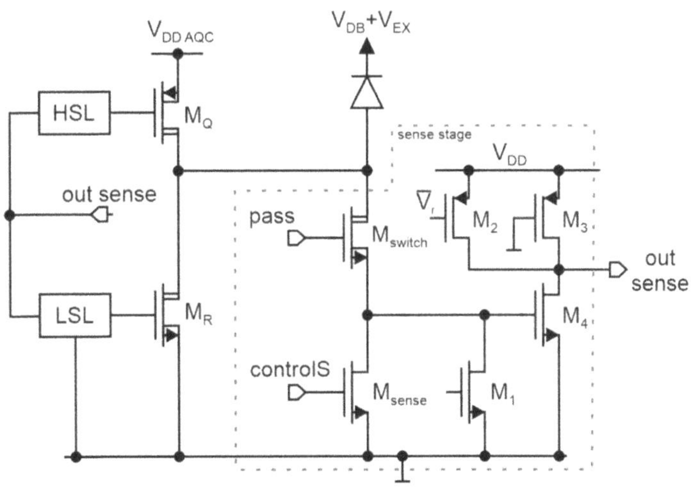

Figure 2.20. Schematic of the quenching circuit presented in [49].

for the LSL. The signals between the low-side and high-side blocks are routed via voltage translators. Experiments were performed using a 100 μm-diameter SPAD bonded to the quenching circuit. The SPAD is described in detail in 1.2.1 [50] (see also figure 1.2.1). A PDP of 40% was reported for a wavelength of 800 nm. This SPAD is manufactured using a custom technology with a thick epitaxial layer optimized for a high PDP at near-infrared (NIR) light. The breakdown voltages of these devices are in the range of 45–55 V [50]. Using an excess bias voltage of 20 V, a DCR of approximately 580 cps was measured for a 50 μm-diameter SPAD. The larger device used for the experiments described in [49] resulted in a higher DCR, but the excess bias voltage was 5 V and lower excess bias voltage reduces the DCR again. The detailed values were not given, but it was noted that the excess bias of 5 V was the best trade-off between the DCR and the PDE. A minimum dead time of 12.2 ns was measured for an excess bias voltage of 5 V. This corresponds to a maximum count rate of 82 Mcps.

In [51], a maximum count rate of 100 Mcps is reported, again for an excess bias voltage of 5 V, due to the best behaviour of the chosen SPAD in terms of PDP and DCR. The circuit uses the same structure as that reported in [49]; the higher count rate is possible due to the shorter reset phase of the SPAD allowed by adjustments in the programmable delay circuit depicted in figure 2.21. A minimum dead time of 10 ns is achieved with a 50 μm-diameter thin SPAD in custom technology bonded to a PCB along with a quenching circuit in 0.18 μm HV CMOS technology from ams AG. Keeping the overall capacitance (consisting of the capacitances of the SPAD, the connection, and the input of the quenching circuit) low is mandatory for high count rates. The APP achieved is 1.8% for the minimum dead time. The maximum excess bias voltage is again 50 V.

The same AQC as that presented in [51] was also used in [52], but with the difference that the AQC was mounted together with the SPAD on a Peltier cooler in a transistor outline (TO) package in a nitrogen atmosphere and was cooled down to −20 °C. The additional cooling increased the performance of the circuit from the 10ns dead time reported above [51] down to 8.3 ns. The excess bias voltage was 5 V, the PDP was 49% at 550 nm, and the DCR was only 9 cps due to the −20 °C temperature. The APP was reported to be 2%. The maximum count rate was 120 Mcount/s.

The single-photon detector presented in [53] used a 0.18μm HV CMOS technology. The circular SPAD consisted of a p+/deep n-well junction and had a diameter of 8μm and a breakdown voltage of 20.3 V. Figure 2.22 depicts the

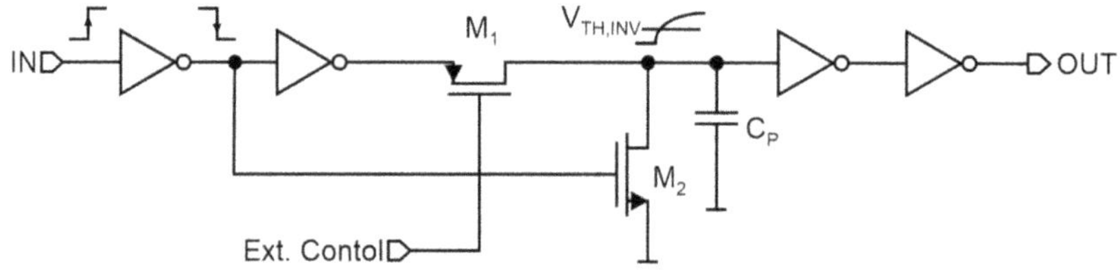

Figure 2.21. Programmable delay stage [51].

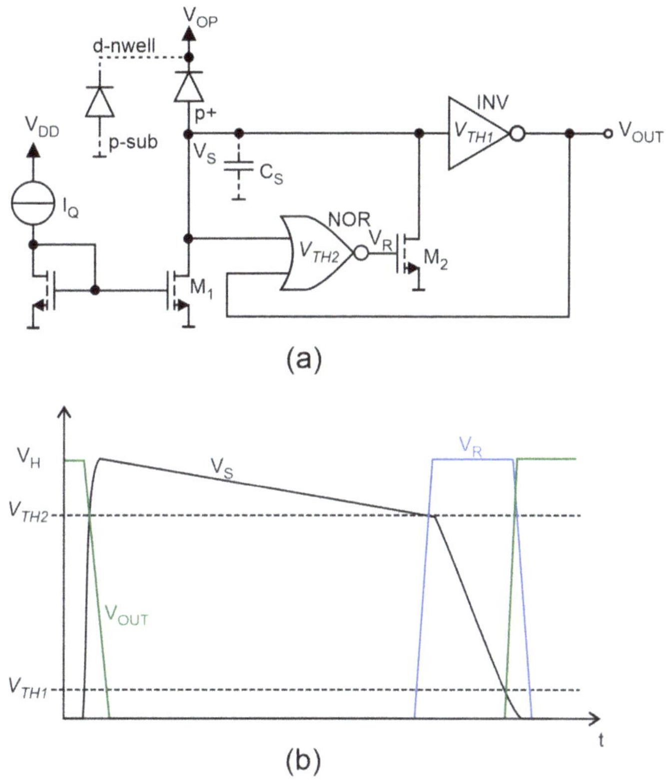

Figure 2.22. Quenching and recharging circuit with dual threshold: (a) principal schematic and (b) voltage diagram [53].

implemented quenching and recharging circuit. When a photon is detected, the avalanche current rapidly increases the voltage V_S to V_H. V_S crosses both thresholds; the first threshold voltage V_{TH1} activates the inverter, and the inverter output goes to ground, which is connected to one input of the NOR gate. M_1 starts to discharge the parasitic capacitance of the input node C_S, V_S starts to decrease until it is less than V_{TH2}, and the NOR gate toggles, opening M_2, and V_S is rapidly pulled towards ground. The hold-off time can be adjusted by the quench current I_Q. The maximum quenching voltage was reported to be 3.5 V, which led to a DCR of 180 cps; afterpulsing effects were not detected. A maximum count rate of 66 Mcps was measured.

In [54], an AQC was presented that used an 150nm HV CMOS technology with lateral double-diffused MOSFETs (LDMOS). The LDMOSs, with a nominal drain–source voltage of up to 40 V, enabled the circuit to operate with high excess bias voltages of up to 35 V. External SPADs with capacitances up to 5 pF could be handled. The circuit offered an improvement compared to [49] by using an n-channel LDMOS transistor instead of a p-channel transistor as the quenching transistor, since it was connected to the cathode of the SPAD instead of the anode.

To obtain the maximum speed from the control circuitry, 1.8 V transistors were used for the HSL as well as in the sense stage, and hence had to be isolated from the substrate by a deep n-well, see figure 2.23.

The voltage regime for the HSL was therefore from V_{DD}, which was 35 V, to $V_{DD, HS}$, which was 1.8 V lower. The LSL was designed using 3.3 V transistors to drive the HV LDMOS M_Q, whose nominal gate voltage was 3.3 V, in the voltage regime from GND to 3.3 V. To connect the HSL and the LSL, a level shifter was necessary to combine the circuits in the different voltage regimes. Figure 2.24 shows the level shifter, which operated between $V_{DD,HS}$ and GND and contained two

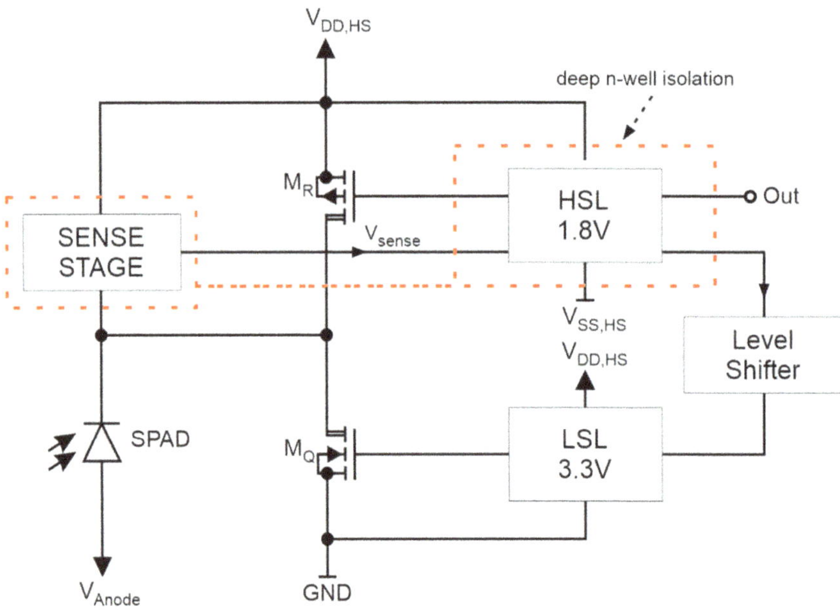

Figure 2.23. Circuit structure described in [54].

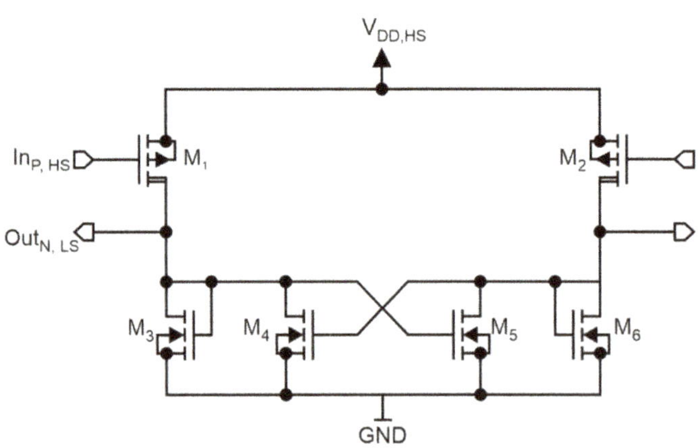

Figure 2.24. Detailed schematic of the level shifter presented in [54].

p-channel LDMOSs which were controlled by the HSL and delivered a differential output for the LSL.

Due to the high excess bias voltages, the avalanche charge has high potential and therefore fast quenching is necessary to keep the avalanche charge as low as possible, which reduces the APP. The presented design achieves a quenching time of 2.2 ns. A reset time of 9.6 ns was accepted to achieve the minimum quenching time. The overall dead time could be adjusted by an additional delay, which could be varied between 5 and 160 ns, leading to a maximum count rate of 90 Mcps for an excess bias voltage of 10 V.

Acconcia *et al* presented an SPAD in [55] that used the transimpedance amplifier (TIA) pickup circuit depicted in figure 2.25. When the SPAD is biased at a voltage higher than the breakdown voltage, transistors M_4 and M_5 are turned off and the circuit is sensitive to incoming photons. HV_1 and HV_2 are cascode transistors that protect the circuitry from HVs. The active input stage consists of M_1 to M_3 as well as R_1 and forms a current input with negative feedback. When a hit occurs, the avalanche current flows through M_3 and mirrors the current towards M_6, generating an output signal which is picked up by a comparator followed by a logic circuit. The SPAD anode is furthermore connected to a set-and-reset HV transistor which is capable of 50V between its gate and drain. These transistors are driven by logic circuits similar to those shown in figure 2.20, but a resistor is added between the quenching and the resetting transistor to provide the necessary pre-bias voltage of approximately 900 mV for the correct functioning of the idle state of the TIA circuit, which the authors reported to be 'rather delicate' [55]. The active quenching circuit was designed using a 180 nm HV CMOS technology and was connected to a SPAD with a diameter of 80 μm produced using a custom technology. The maximum excess bias was 50 V and the the overall dead time was 12.5 ns, corresponding to a maximum count rate of 80 Mcps.

A quadruple voltage-quenching circuit for external SPADs that used a 150nm CMOS technology was presented in [56]. The maximum excess bias voltage was 7.2V, which made it necessary to use four cascode transistors to implement the

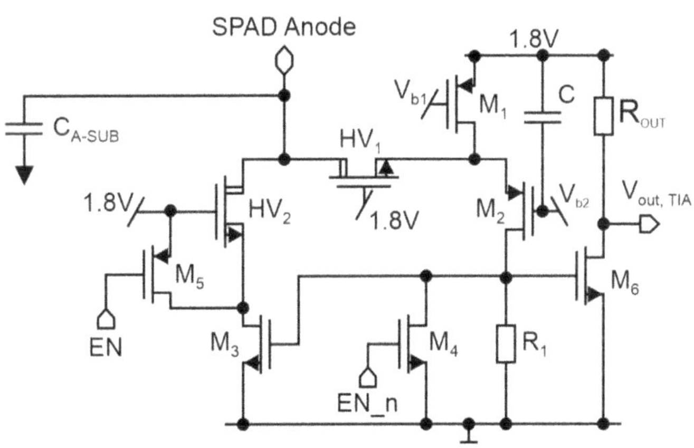

Figure 2.25. Transimpedance pickup circuit [55].

standard 1.8 V devices of this technology. This circuit offers a mixed quenching and active resetting behaviour for external SPADs with capacitances of up to 4 pF. The challenge of this circuit is to operate all the transistors in their safe operating ranges. Figure 2.26 shows a block diagram of the circuit. The logic circuitry is all in the 1.8 V-to-ground voltage regime. In the idle state, the SPAD's cathode is at 1.8 V and as soon as a photon generates an avalanche, the upper part of the switch acts as passive quencher due to an active load. The comparator senses the voltage drop and activates the active quenching pulse logic. The pulse is transformed by a level shifter down to the low-voltage regime between −3.6 and −5.4 V to guarantee a full quenching voltage swing on the SPAD. The switching itself is performed by a quadruple switch whose schematic is shown in figure 2.27; the active load M_{B2} can also be seen in this figure. It is important that there are four cascode transistors between the cathode voltage and each of the switches, M_{S10} for quenching and M_{S1} for reset; this is required for the circuit to operate safely within technology limits.

The post-layout simulations show a minimum dead time of 7.01 ns, which corresponds to a maximum count rate of 142 Mcps.

A maximum count rate of 185 Mcps was reported in [57]. This was reached using a 130nm CMOS technology with an active quenching circuit and an integrated circular SPAD with a diameter of 8 μm. The circuit is depicted in figure 2.28. The avalanche current was first passively quenched by M_3, the voltage drop across M_3 opened M_1, and M_4 pulled the SPAD voltage to a lower value than the breakdown voltage. The reset phase was defined by the inverter chain, which set the dead time; M_3 recharged the SPAD voltage back to V_{ex}. M_3 must be large enough to charge the SPAD, despite incoming photons, to keep the chosen dead time constant. This feature also offered a high dynamic range, which was given as 139 dB. The excess

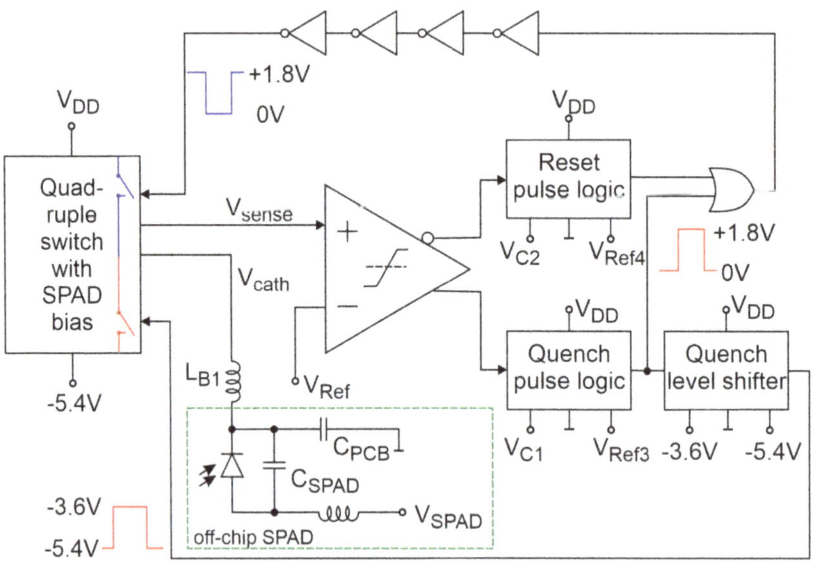

Figure 2.26. Block diagram of the four-cascoder mixed quenching circuit presented in [56].

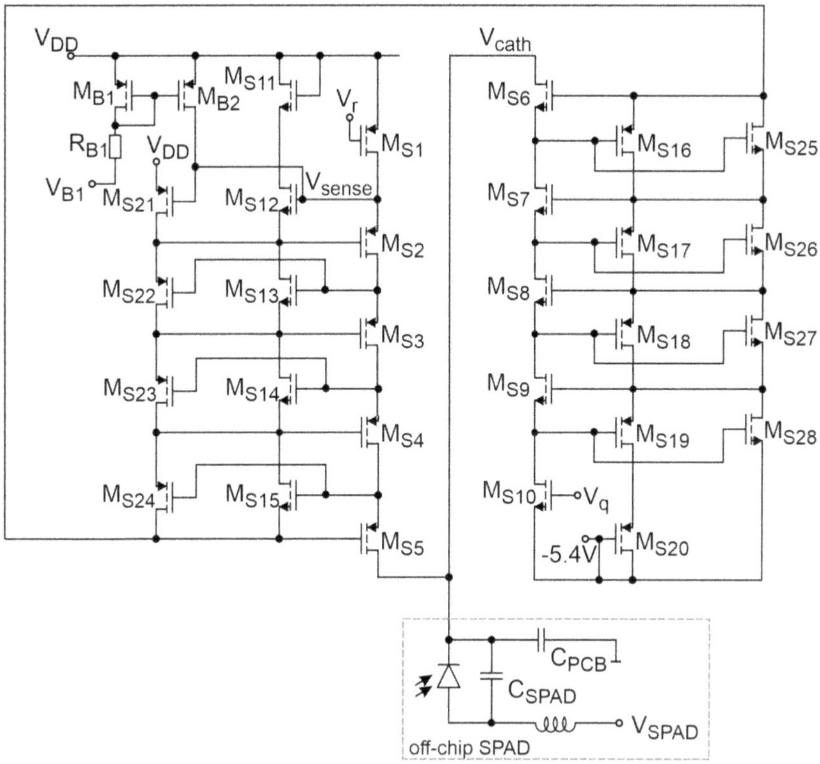

Figure 2.27. Schematic of the quadruple switch and SPAD bias circuit presented in [56].

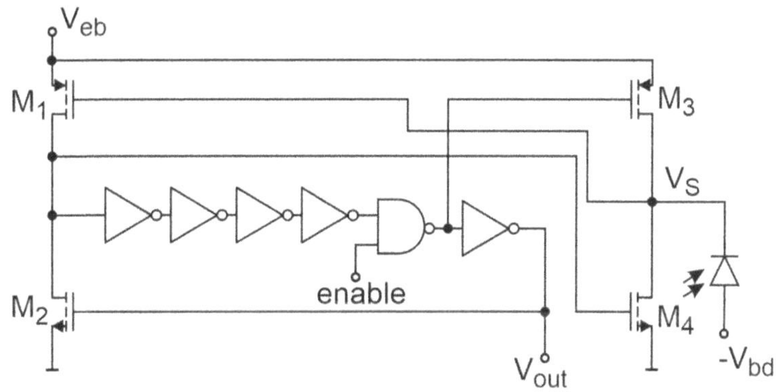

Figure 2.28. Active quenching circuit with a high dynamic range [57].

bias voltage was 2.6 V the minimum dead time was 5.4 ns; for this operating point, the dark count rate was 410 counts/s and the APP was 1.28% for the 8 μm-diameter circular integrated SPAD.

An array of 3 × 3 SPADs integrated together with a reverse bias voltage generation was presented in [58]. Each pixel consists of 3 transistors and the integrated SPAD only. A schematic of the array is shown in figure 2.29. The 90nm CMOS image sensor technology offers 2.5 V transistors, which were used for the circuitry. M_Q is a pre-biased quenching transistor used for passive quenching; M_1, combined with a PMOS pull-up transistor, buffers the output of each pixel toward the readout circuit. The choice of the row to be read out is managed by M_2.

The SPAD's active diameter was only 2 μm, which allowed a pixel pitch of only 5 μm. To realize this small pitch, all pixels shared a deep n-well, which provided the reverse bias voltage for the diodes (see figure 2.30). The breakdown voltage was reported to be 10.3 V. The bias voltage of the SPAD was created internally from the 2/5 V supply voltage using a five-stage charge pump. Due to the thin p+ region, the peak PDP moved to a short wavelength; it was reported to be 36% for 400 nm light. The DCR was around 250 counts/s for a 60 ns dead time and an excess bias voltage of 0.5 V. This led to a count rate of 16.7 Mcps for each pixel. The fill factor of 12% was rather high.

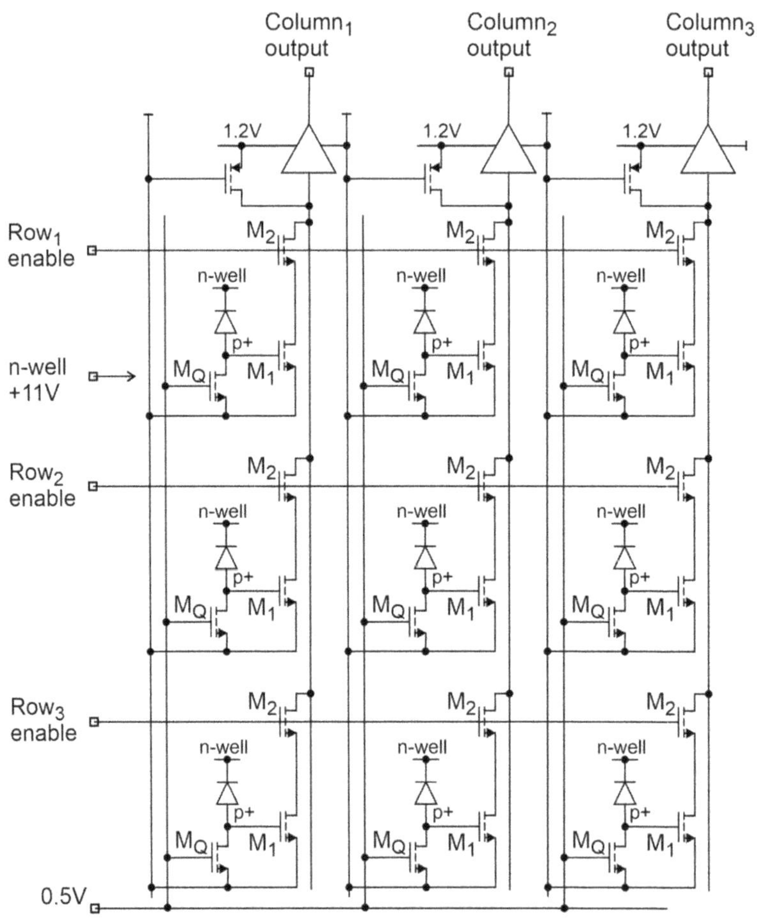

Figure 2.29. Schematic of three-transistor SPAD pixel array [58].

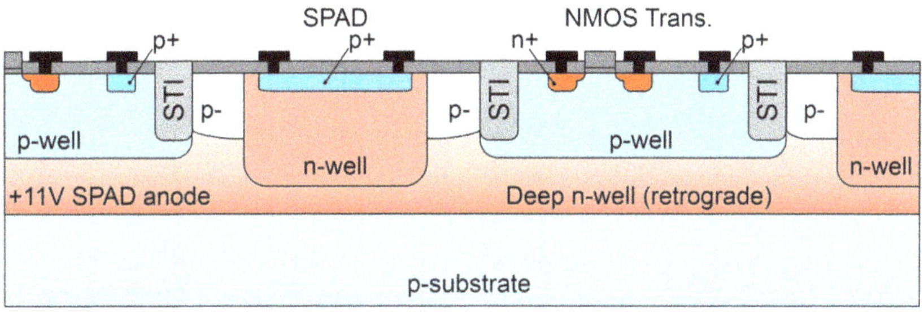

Figure 2.30. Well structure of a three-transistor SPAD pixel array [58].

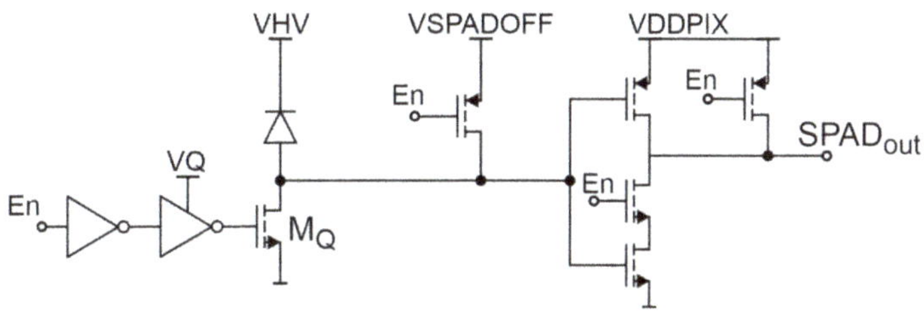

Figure 2.31. Quenching and recharging circuit [59].

Reference [59] presented a fully industrialized SPAD pixel manufactured by STMicroelectronics (STM) using the IMG175 image process [60]. This process has a 130nm front end and a 90 nm back end. The achievable excess bias voltage was given as 0.8–3.3 V, with a breakdown voltage of around 13 V. The median DCR was around 1 kcps at 60 °C and the peak PDP was about 28% for 500 nm light and was still 16.1% and 3.1% at 650 and 850 nm, respectively. The circuit used is depicted in figure 2.31. The incoming avalanche is quenched via M_Q, which is adjusted by V_Q, and the voltage drop is sensed by the inverter. The inverter threshold is adjusted by the supply voltage of the inverter stage V_{DDPIX}. The maximum count rate given was 37 Mcps.

The same authors [61] presented an industrialised SPAD in STM's 40nm node, based on the IMG175 image process mentioned above and using the same circuit. The breakdown voltage was still low, at about 14.6 V, while the excess bias voltage was 1 V. The DCR was less than 700 cps at 60 °C. The fill factor was improved immensely by the addition of microlenses on top of the diodes, and therefore a PDP of more than 70% was reached, compared to 6% in the case of the pixel in the IMG175 technology [59]. The big advantage of the smaller technology node is its lower power consumption, which was given in [62] as a reduction of 85% for a higher maximum count rate of 150 Mcps.

Table 2.2 shows an overview of the described works. The excess biases range from 0.5 to 68 V. Count rates of up to 250 Mcps have been reported. Starting from 0.8 μm CMOS, counting modules as well as quenching and recharging circuits have also been investigated in more modern technologies using CMOS with feature sizes as

Table 2.2. Overview of PCMs as well as active quenching and resetting circuits using integrated circuits.

Ref.	Max. count rate* (Mcps)	Max. V_{EX} (V)	V_{BR}(V)	Int. SPAD (y/n)	Technology	Array
[31]	20	20	439	n	0.8 μm HV CMOS	n
[33]	30	10	16	y (thin)	0.8 μm HV CMOS	n
[34]	100	2.5		y (thin)	0.8 μm CMOS	n
[63]	218	3.3	70	y (thick)	0.35 μm HV CMOS	n
[64]	30	9.9	71.3	y (thick)	0.35 μm HV CMOS	n
[35]	50	6	25	y (thin)	0.35 μm CMOS	n
[37]	35.2	3	2	n	0.35 μm CMOS	n
[41]	25	6	25	y (thin)	0.35 μm HV CMOS	n
[42]	60	6		n	0.18 μm HV CMOS	y (32 ch.)
[44]	33	6		n	0.18 μm HV CMOS	y (64 ch.)
[45]	31	68	339.5, 355	n	0.35 μm HV BCD	n
[47]	250	1.8	20	y	0.18 μm CMOS image sensor	n
[48]	200	3.5	15.5	y (thin)	0.18 μm CMOS	n
[49]	82	50		n	0.18 μm HV CMOS	n
[51]	100	5		n	0.18 μm HV CMOS	n
[52]	120**	5		n	0.18 μm HV CMOS	n
[53]	66	3.5	20.3	y (thin)	180 nm HV CMOS	n
[54]	90	10		n	150 nm HV CMOS	n
[55]	80	50		n	180 nm HV CMOS	n
[56]	142	7.2		n	150 nm HV CMOS	n
[57]	185	2.6		y	130 nm CMOS	n
[58]	16.7	0.5	10.3	y (thin)	90 nm CMOS	y (9 ch.)
[59]	37	3	13	y (thin)	IMG175 (STM)	n
[61]	150	1	14.6	y (thin)	40 nm CMOS	n

*In the case of multi-channel sensors, the figure is per pixel.
**Cooled to −20°C.

low as 40 nm. HV technologies are mainly used for single-photon detection in quantum applications, whereas deep-submicrometer and nanometer CMOS technologies are applied for image sensors with SPADs.

References

[1] Savuskan V, Brouk I, Javitt M and Nemirovsky Y 2013 An estimation of single photon avalanche diode (SPAD) photon detection efficiency (PDE) nonuniformity *IEEE Sensors J.* **13** 1637–40

[2] Mita R and Palumbo G 2008 High-speed and compact quenching circuit for single-photon avalanche diodes *IEEE Trans. Instrum. Meas.* **57** 543–7

[3] Zimmermann H, Steindl B, Hofbauer M and Enne R 2017 Integrated fiber optical receiver reducing the gap to the quantum limit *Sci. Rep.* **7** 2652

[4] Pancheri L, Massari N and Stoppa D 2013 SPAD image sensor with analog counting pixel for time-resolved fluorescence detection *IEEE Trans. Electron Dev.* **60** 3442–9
[5] Dutton N A W, Gyongy I, Parmesan L, Gnecchi S, Calder N, Rae B R, Pellegrini S, Grant L A and Henderson R K 2016 A SPAD-based QVGA image sensor for single-photon counting and quanta imaging *IEEE Trans. Electron Dev.* **63** 189–96
[6] Lee M-J, Ximenes A R, Padmanabhan P, Wang T-J, Huang K C, Yamashita Y, Yaung D-N and Charbon E 2018 High-performance back-illuminated three-dimensional stacked single-photon avalanche diode implemented in 45-nm CMOS technology *IEEE J. Sel. Topics Quant. Electron.* **24** 3801809
[7] Gyongy I, Calder N, Davies A, Dutton N A W, Duncan R R, Rickman C, Dalgarno P and Henderson R K 2018 A 256×256, 100-kfps, 61% fill-factor SPAD image sensor for time-resolved microscopy applications *IEEE Trans. Electron Dev.* **65** 547–54
[8] Kosman J, Almer O, Abbas T A, Dutton N, Walker R, Videv S, Moore K, Haas H and Henderson R 2019 A 500 Mb/s -46.1 dBm CMOS SPAD receiver for laser diode visible-light communication *2019 IEEE Int. Solid-State Circuits Conf. (ISSCC)* pp 468–70
[9] Ximenes A R, Padmanabhan P, Lee M-J, Yamashita Y, Yaung D-N and Charbon E 2019 A modular, direct time-of-flight depth sensor 45/65-nm 3-D-stacked CMOS technology *IEEE J. Solid-State Circ.* **54** 3203–14
[10] Liu M, Hu C, Campbell J C, Pan Z and Tashima M M 2008 Reduce afterpulsing of single photon avalanche diodes using passive quenching with active reset *IEEE J. Quantum Electron.* **44** 430–4
[11] Niclass C, Favi C, Kluter T, Gersbach M and E Charbon 2008 A 128 × 128 single-photon image sensor with column-level 10-bit time-to-digital converter array *IEEE J. Solid-State Circ.* **43** 2977–89
[12] Niclass C, Soga M, Matsubara H, Kato S and Kagami 2013 A 100-m range 10-frame/s 340 × 96-pixel time-of-flight depth sensor in 0.18-μm cmos *IEEE J. Solid-State Circ.* **48** 559–72
[13] Niclass C, Soga M, Matsubara H, Ogawa M and Kagami 2014 A 0.18 μm cmos soc for a 100-m range 10-frame/s 200 × 96-pixel time-of-flight depth sensor *IEEE J. Solid-State Circ.* **49** 315–30
[14] Perenzoni M, Perenzoni D and D Stoppa 2017 A 64 × 64-pixel digital silicon photomultiplier direct tof sensor with 100-mphotons/s/pixel background rejection and imaging/altimeter mode with 0.14% precision up to 6 km for spacecraft navigation and landing *IEEE J. Solid-State Circ.* **52** 151–60
[15] Holma J, Nissinen I, Nissinen J and Kostamovaara J 2017 Characterization of the timing homogeneity in a CMOS SPAD array designed for time-gated Raman spectroscopy *IEEE Trans. Instrum. Meas.* **66** 1837–44
[16] Richardson J, Henderson R K and Renshaw D 2007 Dynamic quenching for single photon avalanche diode arrays *Int. Image Sensor Workshop* pp 258–60
[17] Hyvarinen C, Lidgey F J and Haigh D G 1990 *Analog IC design: the current-mode approach* (Stevenage: Peregrinus)
[18] http://www.hamamatsu.com/eu/en/index.html
[19] http://www.hamamatsu.com/resources/pdf/ssd/c11202series_kacc1207e.pdf
[20] http://www.hamamatsu.com/resources/pdf/ssd/mppc_kapd0006e.pdf
[21] http://www.excelitas.com/
[22] http://www.excelitas.com/product-category/single-photon-counting-modules

[23] http://www.excelitas.com/product/spcm-aqrh
[24] http://www.micro-photon-devices.com/MPD/media/Datasheet/PDM.pdf
[25] http://www.micro-photon-devices.com/home
[26] http://www.micro-photon-devices.com/MPD/media/Datasheet/FastGatedSPAD.pdf
[27] http://www.lasercomponents.com/fileadmin/user_upload/home/Datasheets/lcp/count-series.pdf
[28] http://www.thorlabs.com/newgrouppage9.cfm?objectgroup_id=5255
[29] http://www.thinksrs.com/index.html
[30] http://www.thinksrs.com/downloads/pdfs/catalog/SR400c.pdf
[31] Zappa F, Lotito A, Giudice A C, Cova S and Ghioni M 2003 Monolithic active-quenching and active-reset circuit for single-photon avalanche detectors *IEEE J. Solid-State Circ.* **38** 1298–301
[32] Henri Dautet Pierre, Deschamps Bruno, Dion Andrew D, MacGregor Darleene, MacSween Robert J, McIntyre Claude, Trottier and Paul P 1993 Photon counting techniques with silicon avalanche photodiodes *Appl. Opt.* **32** 3894–900
[33] Zappa F, Tisa S, Gulinatti A, Gallivanoni A and Cova S 2004 Monolithic CMOS detector module for photon counting and picosecond timing *30th European Solid-State Circuits Conference (IEEE Cat. No.04EX850)* pp 341–4
[34] Rochas A, Besse P-A and Popovic R S 2004 Actively recharged single photon counting avalanche photodiode integrated in an industrial CMOS process *Sensors Actuat. A: Physical* **110** 124–9 Selected Papers from Eurosensors XVI Prague, Czech Republic
[35] Bronzi D, Tisa S, Villa F, Bellisai S, Tosi A and Zappa F 2013 Fast sensing and quenching of CMOS SPADs for minimal afterpulsing effects. *IEEE Photon. Technol. Lett.* **25** 776–9
[36] Bronzi Danilo, Villa Federica, Bellisai Simone, Markovic Bojan, Tisa Simone, Tosi Alberto, Zappa Franco, Weyers Sascha, Durini Daniel, Brockherde Werner and Paschen Uwe 2012 Low-noise and large-area CMOS SPADs with timing response free from slow tails. *European Solid-State Device Research Conf. (ESSDERC)* pp 230–3
[37] Deng Shijie and Morrison Alan P 2012 Active quench and reset integrated circuit with novel hold-off time control logic for Geiger-mode avalanche photodiodes *Opt. Lett.* **37** 3876–8
[38] Ming Chen Chenghao, Li Alan P, Morrison Shijie, Deng Chuanxin, Teng Houquan, Liu Hongchang, Deng Xianming and Yuan Libo 2020 Design and implementation of a compact single-photon counting module *Electronics* **9** 1131
[39] Jackson J C, Donnelly J, O'Neill B, Kelleher A-M, Healy G, Morrison A P and Mathewson A 2003 Integrated bulk/SOI APD sensor: bulk substrate inspection with Geiger-mode avalanche photodiodes *Electron. Lett.* **39** 735–6
[40] Deng Shijie, Morrison Alan P, Liu Houquan, Deng Hongchang, Chen Ming, Yuan Libo and Teng Chuanxin 2019 High dynamic range photo-detection module using on-chip dual avalanche photodiodes *IEEE Photon. Technol. Lett.* **31** 1940–3
[41] Villa Federica, Bronzi Danilo, Zou Yu, Scarcella Carmelo, Boso Gianluca, Tisa Simone, Tosi Alberto, Zappa Franco, Durini Daniel, Weyers Sascha, Paschen Uwe and Brockherde Werner 2014 CMOS SPADs with up to 500 μm diameter and 55% detection efficiency at 420 nm *J. Modern Opt.* **61** 102–15
[42] Cuccato A, Antonioli S, Crotti M, Labanca I, Gulinatti A, Rech I and Ghioni M 2013 Complete and compact 32-channel system for time-correlated single-photon counting measurements *IEEE Photonics* J. **5** 6801514
[43] Cammi C, Gulinatti A, Rech I, Panzeri F and Ghioni M 2011 Compact eight channel SPAD module for photon timing applications *Proc. SPIE* **8033** 80330H

[44] Ceccarelli F, Gulinatti A, Labanca I, Rech I and Ghioni M 2016 Gigacount/second photon detection module based on an 8 × 8 single-photon avalanche diode array *IEEE Photon. Technol. Lett.* **28** 1002–5

[45] Fang Yu-Qiang, Luo Kai, Gao Xing-Guo, Huo Gai-Qing, Zhong Ang, Liao Peng-Fei, Pu Pu, Bao Xiao-Hui, Chen Yu-Ao, Zhang Jun and Pan Jian-Wei 2020 High detection efficiency silicon single-photon detector with a monolithic integrated circuit of active quenching and active reset *Rev. Sci. Instrum.* **91** 123106

[46] http://www.excelitas.com/de/product-category/single-photon-counting-modules

[47] Katz A, Blank T, Fenigstein A, Leitner T and Nemirovsky Y 2019 Active-reset for the n +p single-ended SPAD used in the NIR LiDAR receivers *IEEE Trans. Electron Dev.* **66** 5191–5

[48] Xu Y, Lu J and Wu Z 2020 A compact high-speed active quenching and recharging circuit for SPAD detectors *IEEE Photonics* J. **12** 1–8

[49] Acconcia Giulia, Rech Ivan, Gulinatti Angelo and Ghioni Massimo Aug 2016 High-voltage integrated active quenching circuit for single photon count rate up to 80 Mcounts/s *Opt. Express* **24** 17819–31

[50] Gulinatti A, Rech I, Panzeri F, Cammi C, Maccagnani P, Ghioni M and Cova S 2012 New silicon SPAD technology for enhanced red-sensitivity, high-resolution timing and system integration *J. Modern Opt.* **59** 1489–99

[51] Acconcia G, Labanca I, Rech I, Gulinatti A and Ghioni M 2017 Note: Fully integrated active quenching circuit achieving 100 MHz count rate with custom technology single photon avalanche diodes *Rev. Sci. Instrum.* **88** 026103

[52] Ceccarelli F, Acconcia G, Labanca I, Gulinatti A, Ghioni M and Rech I 2018 152-db dynamic range with a large-area custom-technology single-photon avalanche diode *IEEE Photonics Technol. Lett.* **30** 391–4

[53] Niclass C and Soga M 2010 A miniature actively recharged single-photon detector free of afterpulsing effects with 6 ns dead time in a $0.18\mu m$ CMOS technology *2010 Int. Electron Devices Meeting* 14.3.1–4

[54] Jungwirth M, Dervić A and Zimmermann H 2020 Integrated high voltage active quenching circuit in 150 nm CMOS technology *2020 Austrochip Workshop on Microelectronics (Austrochip)* pp 53–6

[55] Acconcia G, Ghioni M and Rech I Dec 2018 37 ps-precision time-resolving active quenching circuit for high-performance single photon avalanche diodes *IEEE Photon.* J. **10** 1–13

[56] Dervić A, Goll B and Zimmermann H 2020 Quadruple voltage mixed quenching and active resetting circuit in 150 nm CMOS for an external SPAD *2020 23rd Int. Symp. on Design and Diagnostics of Electronic Circuits Systems (DDECS)* pp 1–5

[57] Eisele A, Henderson R, Schmidtke B, Funk T, Grant L, Richardson J and Freude W 2011 185 MHz count rate, 139 dB dynamic range single-photon avalanche diode with active quenching circuit in 130 nm CMOS technology *Int. Image Sensor Workshop (IISW'11), (Hokkaido, Japan, June 8–11, 2011)* R43.

[58] Henderson R K, Webster E A G, Walker R, Richardson J A and Grant L A 2010 A 3×3, $5\mu m$ pitch, 3-transistor single photon avalanche diode array with integrated 11 V bias generation in 90 nm CMOS technology *2010 Int. Electron Devices Meet.* 14.2.1–4

[59] Pellegrini S and Rae B 2017 Fully industrialised single photon avalanche diodes *Proc. SPIE* **10212** 102120D

[60] http://www.st.com/content/st_com/en/about/innovation–technology/imaging.html

[61] Pellegrini S, Rae B, Pingault A, Golanski D, Jouan S, Lapeyre C and Mamdy B 2017 Industrialised SPAD in 40 nm technology *2017 IEEE Int. Electron Devices Meet. (IEDM)* 16.5.1–4

[62] https://mycmp.fr/wp-content/uploads/2021/04/2019_cmp_usersmeeting_04_sarapellegrini_cmp_-spad_2019-02-06.pdf.

[63] Dervić A, Goll B, Steindl B and Zimmermann H 2019 Transient measurements and mixed quenching, active resetting circuit for SPAD in $0.35\ \mu\mathrm{m}$ high-voltage CMOS for achieving 218 Mcps *2019 26th IEEE Int. Conf. on Electronics, Circuits and Systems (ICECS)* pp 819–22

[64] Dervić Alija, Steindl Bernhard, Hofbauer Michael and Zimmermann Horst 2019 High-voltage active quenching and resetting circuit for SPADs in 0.35 μm CMOS for raising the photon detection probability *Opt. Eng.* **58** 1–4

IOP Publishing

Single-photon Detection for Data Communication and Quantum Systems

Michael Hofbauer, Kerstin Schneider-Hornstein and Horst Zimmermann

Chapter 3

Advanced quenching and gating of integrated SPADs

In this chapter, we introduce advanced active quenching circuits that increase the excess bias voltage of single-photon avalanche diodes (SPADs) from a single supply voltage to twice, three times, or even four times the usual circuit supply voltage, i.e. to 13.2 V in 0.35 μm complementary metal–oxide–semiconductor (CMOS) with a nominal supply voltage of 3.3 V. The higher excess bias voltages of SPADs lead to enhanced photon detection probabilities. The advanced active quenching circuits combine a low avalanche detection threshold and fast quenching to keep after-pulsing probabilities low at high photon detection probabilities. In addition, the principle of gating SPADs is introduced and an advanced gating circuit that exploits cascoding to double the SPAD excess bias from 3.3 to 6.6 V is described.

3.1 Advanced quenching

3.1.1 Single-supply-voltage quenching circuit

A mixed quenching active reset circuit was described in [1]. This circuit was realized in 0.35μm high-voltage CMOS together with the high-voltage (HV) CMOS SPAD introduced in section 1.3.2 on the same chip. The circuit diagram of this quenching circuit is shown in figure 3.1. The SPAD has a diameter of 90 μm and a capacitance of 150 fF. A square minipad with a side length of 35 μm (with a capacitance of about 25 fF) was implemented at the cathode node to be able to measure the quenching and resetting transients.

The current mirror M_{B1}–M_{B2} biases the SPAD to V_{DD} to address leakage currents, which tend to discharge it, e.g. during long intervals between dark counts or photon detections. In the quenching branch, the source follower M_{18} reduces the detection threshold from V_{DD} to the inverter threshold of about $V_{DD}/2$ by the

doi:10.1088/978-0-7503-2584-4ch3

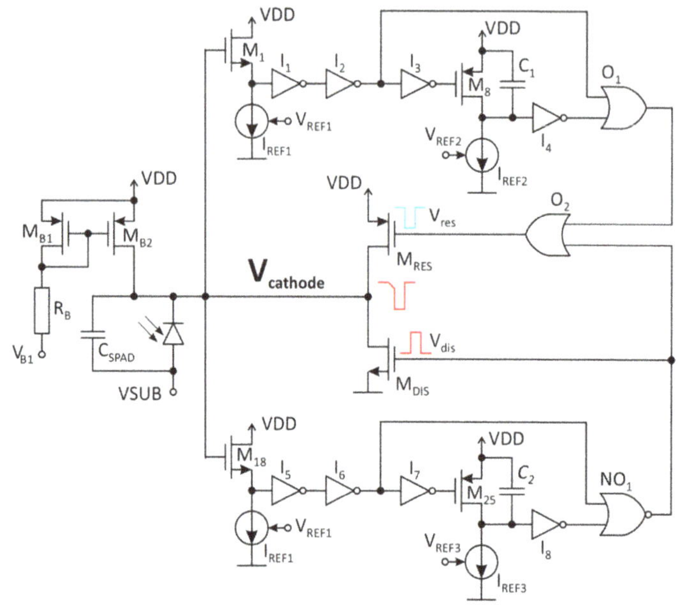

Figure 3.1. Simplified schematic of an active quenching circuit. © 2019 IEEE. Reprinted, with permission, from [1].

gate–source voltage of M_{18}. This gate–source voltage can be changed by varying the current I_{REF1} via V_{REF1} (I_{REF1} is proportional to V_{DD}-V_{REF1}); the gate–source voltage decreases for increasing values of V_{REF1}, meaning that passive quenching takes longer. When the voltage drop across M_{B2} due to the increasing avalanche current is large enough for inverter I_5 to switch to 'high' and I_6 to switch to 'low', the NOR gate NO_1 switches to 'high' and the discharging transistor M_{DIS} turns on, actively quenching the SPAD at a voltage less than its breakdown voltage. M_{25} is turned off by I_7 and I_{REF3} charges C_2. After the time t_q determined by V_{REF3} and C_2, I_8 switches to 'high' and NO_1 switches to 'low,' turning off M_{DIS}, i.e. it finishes the active quenching time. M_{RES} is off until C_1 is charged below the inverter threshold of I_4. This time is longer than t_q, and the difference is equal to the duration t_r of the reset pulse. It is possible to set t_q and t_r.

A chip photo of the mixed quenching active-resetting circuit is shown in figure 3.2. This circuit occupied an active area of 100 μm × 125 μm. The transient measurements at the cathode node were performed using a 3 GHz picoprobe that had an input capacitance of 100 fF, an input resistance of 10 MΩ, and an attenuation of 1:20. Therefore, the total capacitance at the cathode node was 275 fF (SPAD, minipad, and picoprobe). The transients are shown in figure 3.3, in which the actual measured voltage values were multiplied by 20 to compensate for the attenuation of the picoprobe. It is clearly visible that the passive quenching phase lasts for about 1.2 ns for V_{REF1} = 2.23 V, about 3.7 ns for V_{REF1} = 2.72 V, and about 5.1 ns for V_{REF1} = 3.22 V. Thus, increasing the gate–source voltage of M_{18} (and M_1) leads to earlier active quenching. Considering the passive quenching phase of about 5 ns for

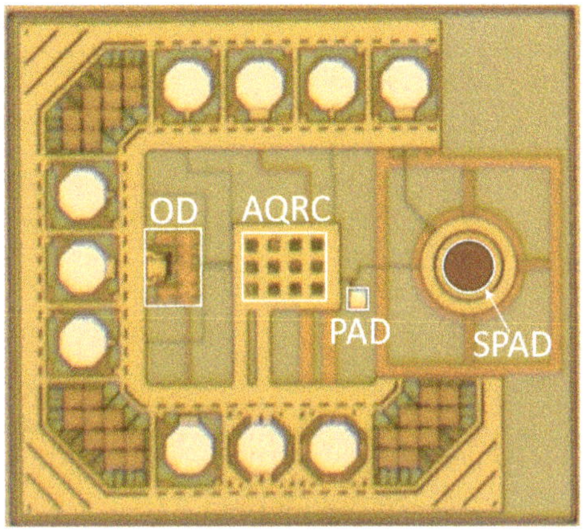

Figure 3.2. Chip photo of the active quenching circuit. © 2019 IEEE. Reprinted, with permission, from [1].

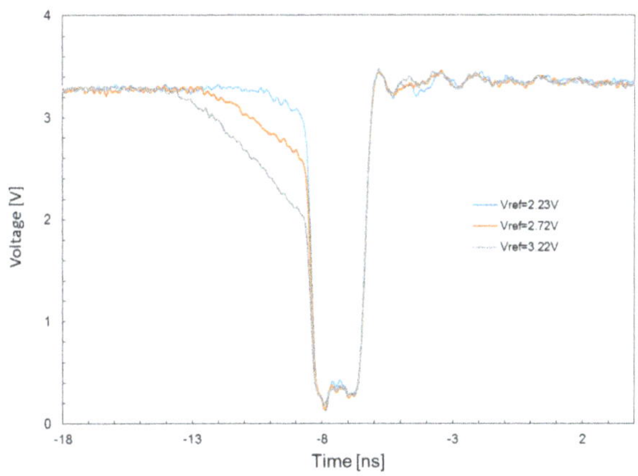

Figure 3.3. Measured voltage transients at the cathode node of the SPAD [1].

V_{REF1} = 3.22 V with a voltage drop of about 1.3 V, it is estimated that the total avalanche build-up time is about 10 ns to 12 ns.

The time taken from the onset of active quenching to the end of quenching was about 0.67 ns (for V_{REF1} = 2.23 V). The rising edge for resetting lasted about 0.63 ns. This 0.67 ns quenching time is somewhat larger than the 0.44 ns reported in [2] for a SPAD capacitance of 60 fF, due to the larger SPAD capacitance and the picoprobe capacitance.

3.1.2 Double-supply-voltage quenching circuit

The excess bias voltage can be raised in order to increase the PDP of SPADs. This leads to a requirement for fast transistors that have high breakdown voltages. High-voltage transistors, however, possess longer gate lengths in order to realise higher breakdown voltages, which makes them slower. Fortunately a circuit technique exists that uses fast low-voltage transistors and doubles the voltage blocking

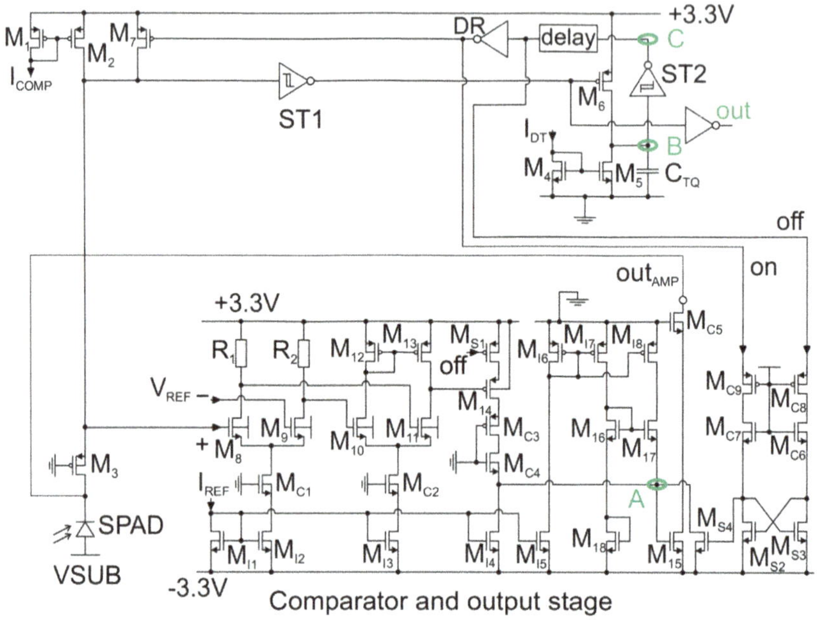

Figure 3.4. Active quenching circuit for excess bias voltages of up to 6.6 V [3].

capability of these fast transistors. This technique is called cascoding. Two transistors are stacked: M_3 and M_7 in figure 3.4 connecting the SPAD to +3.3 V; M3 is the so-called cascode transistor that extends the blocking capability of M_7 from 3.3 to 6.6 V. This allows the SPAD to be quenched to −3.3 V by M_{15} and the cascode transistor MC_5, i.e. a voltage swing of 6.6 V is possible at the SPAD's cathode. When the substrate V_{SUB} is biased to V_{BR}-3.3 V (note that V_{SUB} and V_{BR} are negative), the excess bias voltage of 6.6 V results.

The active quenching circuit shown in figure 3.4 includes a fast comparator (M_8–M_{11}), a cascoded output stage and the cascoded switching stage with M_{15} and MC_5 [3]. M_2 acts as an active resistor. The comparator can detect an avalanche event in its early stages, when the avalanche current from the SPAD causes a voltage drop of only 100 mV across M_2. Thanks to the very short delay of the comparator and the output stage (0.56 ns), the SPAD is completely quenched (the switching time of the output stage is 0.44 ns) in 1.0 ns. This short delay is only possible in 0.35 μm CMOS, because the gate of M_{15} is prebiased via M_{16}–M_{18}. The voltage drop across M_{18} is mirrored to the gate of M_{15} by M_{16} and M_{17}, which results in a certain drain current of M_{15} (through MC_5, M_3, and M_2).

Some logic (M_{S2}–M_{S4}, M_{C6}–M_{C9}, the Schmitt triggers ST1 and ST2) and a delay block control reset (which charges the SPAD to $|VBR|$+VEX by switching on M_7; M_{15} is off) complete the circuit and define the hold-off time (dead time) [3]. In the waiting phase, the SPAD is sensitive and M_7 as well as M_{15} are off. During active quenching, M_7 is off and M_{15} is on. During the first phase of the avalanche build-up, M_2 acts as a kind of passive quenching resistor, until the voltage drop across M_2 reduces the potential at the non-inverting input of the comparator to less than the decision threshold defined by V_{REF} and until the delay time of the comparator and its output stage switches on M_{15}. Therefore, active quenching starts about 0.56 ns after the comparator threshold was reached and is completely finished after another 0.44 ns.

The cascoded quencher integrated together with the thick SPAD in 0.35 μm pin-photodiode CMOS increased the PDP for 635 nm photons from about 22% with a 3.3 V excess bias to 35.1% with a 6.6 V excess bias [3]. The afterpulsing probability (APP) for $V_{EX} = 6.6$ V (see figure 3.5) was reduced from 14.7% with $V_{REF} = 0$ V (3.3 V voltage drop across M2) to 4.8% with $V_{REF} = 3.2$ V (0.1 V voltage drop across M_2). These results impressively show that an active quencher in 0.35 μm CMOS can realise both high PDP and low APP.

3.1.3 Triple-supply-voltage quenching circuits

Figure 3.6 shows the circuit diagram of an active quenching circuit using the high-speed low-voltage transistors of a 0.35 μm high-voltage CMOS technology but allowing excess bias voltages of the SPAD that are up to three times the circuit supply voltage of 3.3 V [4]. The SPAD biasing circuit is similar to that of the double-supply-voltage quencher described above. The comparator COMP is the same as in the double-supply-voltage quencher. The cascoded switches are extended by a third transistor each: M4 increases the voltage blocking capability of the SPAD charging switch (M6, M5, and M4) to 9.9 V, and M3 allows the SPAD discharging switch (M1, M2, and M3) to have a switching capability of 9.9 V. The gates of M4 and M3 need dynamic biasing by M8 and M7, respectively. It should be mentioned that the wells of M1–M8 have to be connected to their corresponding sources, which is possible due to the triple-well CMOS process used. At the output of the comparator, which is supplied by +3.3 V and −3.3 V, a level shifter toward −6.6 V is present. This level shifter consists of M9 to M13 and is able to drive the 9.9 V switch formed by

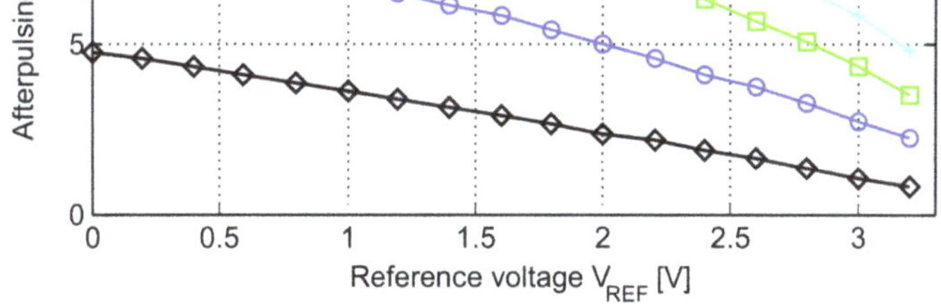

Figure 3.5. Dependence of APP on reference voltage V_{REF}. © 2018 IEEE. Reprinted, with permission, from [3].

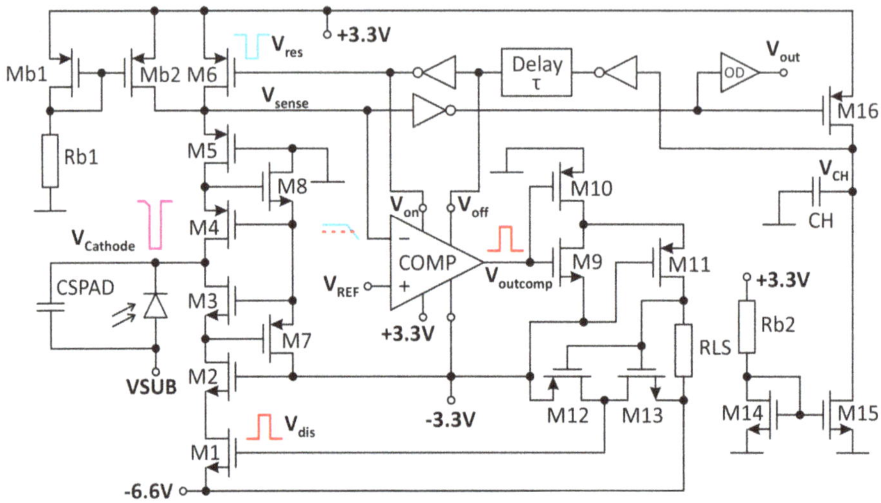

Figure 3.6. Active quenching circuit for excess bias voltages of up to 9.9 V. Reproduced from [4] with permission from SPIE.

M1 to M3. In this way, the cathode of the SPAD can be charged to +3.3 V and left floating (connected by the active resistor Mb2 to 3.3 V) until a photon triggers an avalanche and the comparator (which has a detection threshold of 0.1 V (V_{REF} = 3.2 V)) switches on M1, which quenches the SPAD. After a hold-off time (the dead time), M1 is switched off and M6 is switched on to charge the SPAD again. According to circuit simulations, the power consumption of this quencher in the idle state is about 4.5 mW, which increases to about 10.2 mW for 100 photon detections within 3 μs [4].

The chip was fabricated using a 0.35 μm HV CMOS process with an isolation capability of down to −100 V for the substrate potential, i.e. the anode potential of the integrated SPAD (see figure 1.40). A microphotograph of the triple-supply-voltage quencher chip is shown in figure 3.7. The active area of the chip is 260 μm × 138 μm.

This quencher was used to characterise the HV CMOS SPAD with respect to dark count rate (DCR), afterpulsing probability (APP), and photon detection probability (PDP). These results are presented in figures 3.8, 3.9, and 3.10. The breakdown voltage of this SPAD was 71.3 V at 25 °C (the corresponding substrate potential was therefore −68 V; for V_{EX} = 9.9 V, the substrate potential had to be −77.9 V). The dead time for these measurements was 33 ns. At 9.9 V of excess bias, the DCR was 63 kHz, the APP was 6.7% and the PDP was 67.8% for a wavelength of 642 nm. This PDP compares to 44% at 690 nm [5], 41% at 450nm [6], 55% at 420 nm [7] and 50% at 550nm [8].

In the following, another triple-voltage mixed quenching active-resetting circuit (TVQC) [9] will be introduced, which uses 5 V transistors to reduce the number of transistors in the quenching and resetting paths and in turn to speed up the quenching and resetting. This circuit was realized in the PIN-photodiode 0.35μm CMOS technology with a nominal supply voltage of 3.3 V and the SPAD

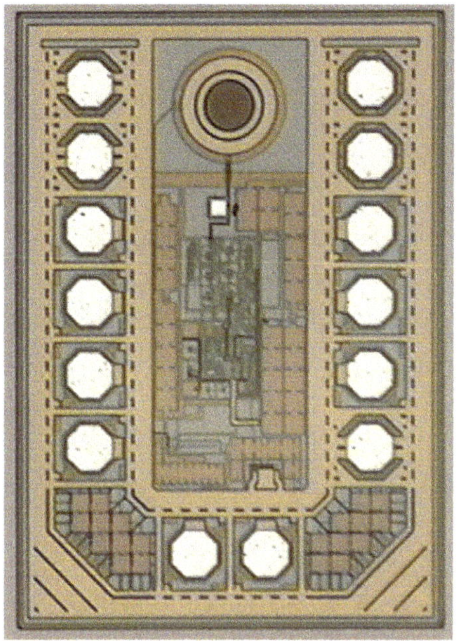

Figure 3.7. Chip photo of the active quenching circuit for excess bias voltages of up to 9.9 V.

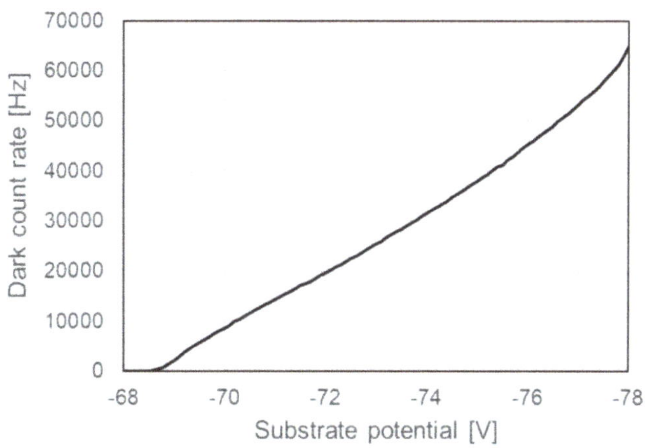

Figure 3.8. Dark count rate for excess bias voltages of up to 9.9 V. Reproduced from [4] with permission from SPIE.

implemented in this technology was described in chapter 1. Figure 3.11 shows the principle of the circuit and figure 3.12 depicts the detailed circuit diagram of the TVQC. The quenching and resetting paths (section A in figure 3.12) use the same cascoding approach as that of figure 3.4 [3], but using 5 V transistors instead of 3.3 V transistors, which increases the excess bias from 6.6 V to 10 V (9.9 V). This results in the use of only four transistors within the quenching (MQ1 and MQ2) and resetting (MR1 and MR2) paths, compared to the eight transistors required in the 9.9 V quencher shown in figure 3.6. Therefore, fewer parasitic capacitances are present, the

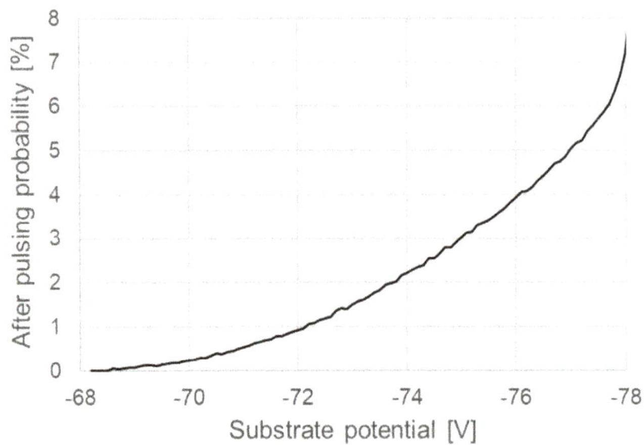

Figure 3.9. Afterpulsing probability for excess bias voltages of up to 9.9 V. Reproduced from [4] with permission from SPIE.

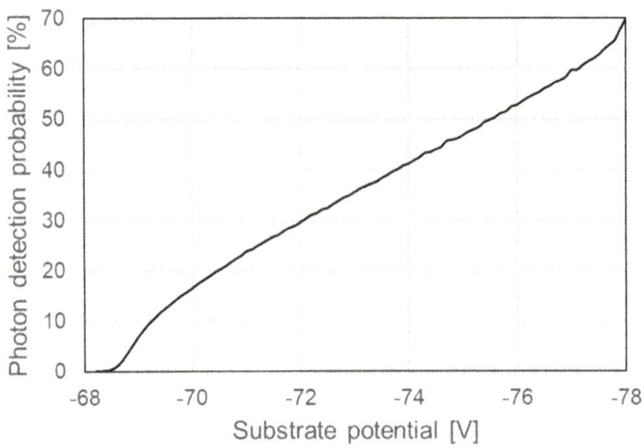

Figure 3.10. Photon detection probability for excess bias voltages of up to 9.9 V. Reproduced from [4] with permission from SPIE.

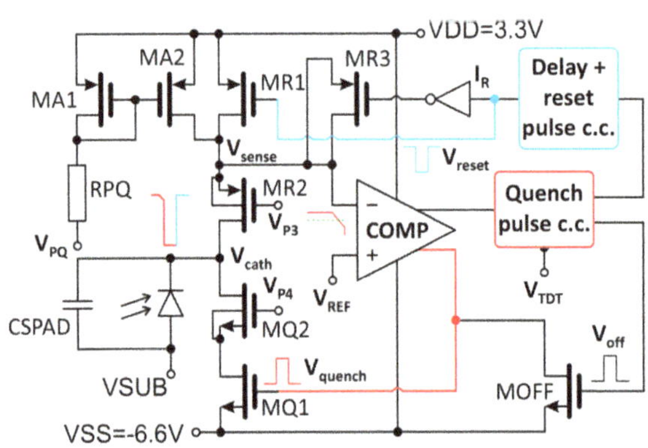

Figure 3.11. Triple-voltage quenching circuit using 5 V transistors for excess bias voltages of up to 9.9 V [9].

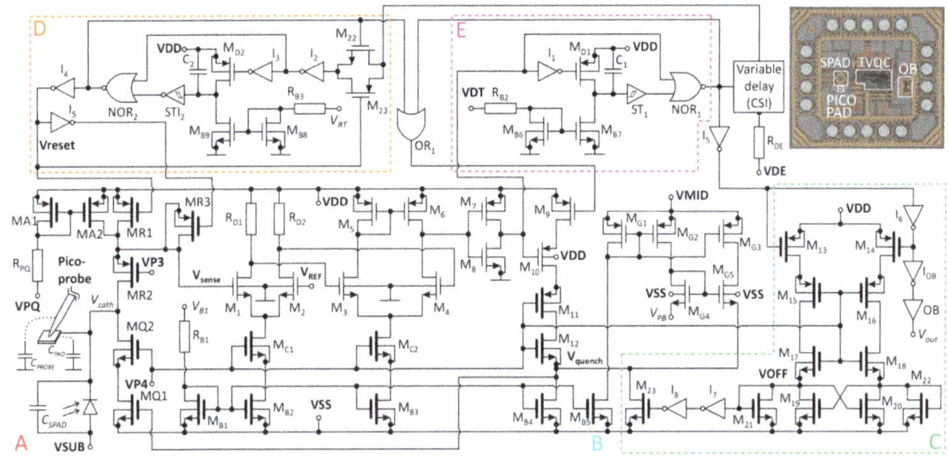

Figure 3.12. Detailed schematic and a chip photo of a triple-voltage quenching circuit using 5 V transistors for excess bias voltages of up to 9.9 V [9].

circuit's complexity is lower, and it needs a smaller chip area. Due to the reduced number of parasitics, faster quenching is achievable and/or larger SPAD capacitances can be handled. To be able to verify the fast quenching, a small pad (a so-called pico-pad) with a size of $20 \times 20 \mu m^2$ was integrated into a high metal layer to measure the transients at the SPAD's cathode node with the help of a picoprobe.

The SPAD integrated together with this TVQC had an active diameter (the diameter of the p-well) of $40\,\mu m$. Its breakdown voltage was 34.2 V, which corresponds to $V_{SUB} = -30.9$ V for $V_{DD} = 3.3$ V to bias the SPAD at its breakdown voltage [9]. To enter the Geiger mode, even more negative substrate potentials are necessary. The CMOS process used is a triple-well CMOS process, which allows the MOSFETs to be isolated from such negative substrate potentials. Section B (figure 3.12) contains a four-stage comparator with pre-bias (MG1–MG5) to speed up the quenching process. Five-volt transistors are necessary in the cascoded constant-current sources of the first and second differential amplifier stages and in the output level shifter to drive MQ1. Section C contains the comparator's disabling circuit and the output buffer OB. Section D depicts the quenching pulse-conditioning circuit, and section E shows the resetting pulse-conditioning circuit, which includes a delay.

The operating principle is shown in figure 3.13 with simulated transients of important circuit nodes. This figure also defines separate time intervals for the circuit operation. It also depicts the different voltage levels of circuit nodes. A photon triggers an avalanche at t_1 and passive quenching starts leading to a slowly decreasing cathode potential V_{cath}. At t_2, the comparator's reference voltage V_{REF} is crossed. At t_3, the comparator's output switches on the quenching MOSFET MQ1, and active quenching starts, which discharges the SPAD's cathode to V_{SS} at t_4. After a hold-off time, at t_5, SPAD charging, i.e. the resetting phase, starts. At t_6, the V_{sense} node reaches V_{REF} again and the resetting phase has to be extended to t_7 to disable the comparator. Finally, at t_7, the next photon can be detected.

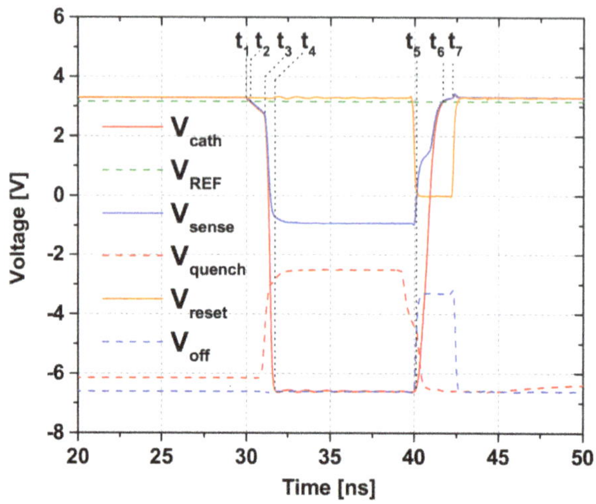

Figure 3.13. Operational principle of a triple-voltage quenching circuit using 5 V transistors for excess bias voltages of up to 9.9 V (V_{mref}=3.2 V, T_{DT}=12.5 ns). © 2021 IEEE. Reprinted, with permission, from [9].

Before we look at the measured results, it should be noted that there is an additional feature in this TVQC, compared to the circuit of figure 3.4. The PMOS transistor MR3 connected as a capacitor compensates for the charge which is injected by the gate–drain overlap capacitance of MR1 during the rising edge of V_{reset} at the end of the charging interval, t_7. This injection charge would make the node V_{sense} more positive than V_{DD}, and the passive quenching phase would then last longer when the next avalanche is triggered by an absorbed photon. MR3 has an opposite edge at its gate due to the inverter I_5, and its gate width can be optimized to obtain the best compensation for the injection charge. The power consumption of the TVQC is 14 mW for the dead time T_{DT} = 7.9 ns, and the chip area of the TVQC is 236 × 108 μm^2 without the buffer, which adds 120 × 86 μm^2 [9].

For the measurements, a chip was kept at 25 °C using a Peltier element. The transient of the cathode potential was measured using a picoprobe 34 A from GGB Industries, which added its input capacitance of 100 fF in parallel to the SPADs capacitance at the pico-pad. The impact of the input resistance of the picoprobe (10 MΩ) was negligible. The 20:1 attenuation of the picoprobe was considered by multiplying the obtained voltages by 20 in figure 3.14. The picoprobe was connected to a Keysight MSOV204A oscilloscope. The measured reaction time $t_3 - t_2$ (during the reaction time, passive quenching continues because of the comparator's delay) is 0.82 ns, which is only slightly longer than the simulated value of 0.78 ns without the picoprobe's input capacitance. The measured pull-down phase ($t_4 - t_3$) lasts for 0.88 ns (0.61 ns simulated without the picoprobe). The measured total quenching time ($t_4 - t_2$) is 1.7 ns compared to a simulated value of 1.39 ns without the picoprobe for the full swing of 9.8 V ($V_{REF} - V_{SS}$). The resulting slew rate of the TVQC is therefore 7.05 GV s^{-1}, which is 1.29, 1.30, and 2.05 times faster than those reported in [3, 10], and [11], respectively. The measured resetting time ($t_6 - t_5$) is 1.75 ns, compared to the value of 1.59 ns simulated without the input capacitance of the picoprobe [9].

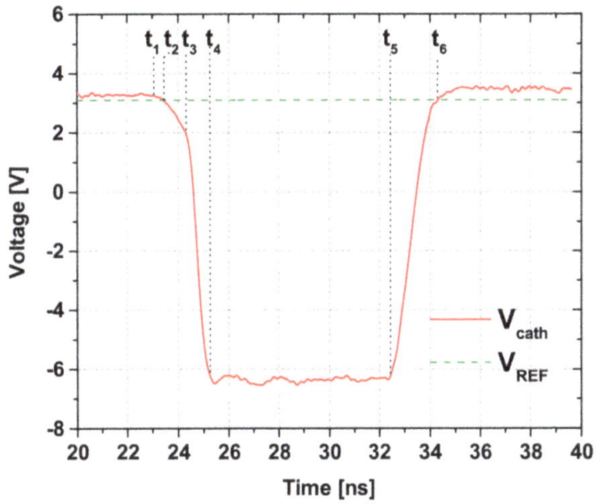

Figure 3.14. Measured transient response of the cathode node of a triple-voltage quenching circuit using 5 V transistors for excess bias voltages of up to 9.9 V ($V_{\rm mref}$=3.2 V, $T_{\rm DT}$=12.5 ns) [9].

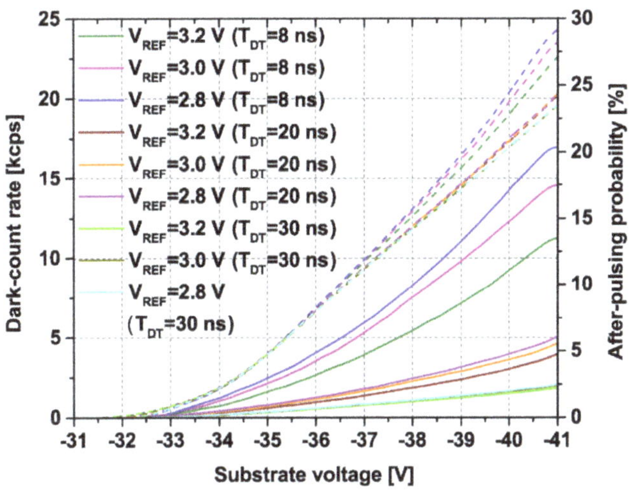

Figure 3.15. Measured DCR (dashed lines) and APP (solid lines) obtained with a triple-voltage quenching circuit using 5 V transistors for excess bias voltages of up to 9.9 V. © 2021 IEEE. Reprinted, with permission, from [9].

An NI PXIe 5162 digitizer was connected to the chip's output and used to characterize the DCR, APP, and PDP. The influences of the excess bias voltage (obtained by varying $V_{\rm SUB}$), the comparator's reference voltage, and the dead time $T_{\rm DT}$ were investigated (see figure 3.15). A maximum DCR of 24.3 kcps is obtained at $V_{\rm EX}$ = 9.9 V, $V_{\rm REF}$ = 3.2 V, i.e. at a detection threshold of 0.1 V, and $T_{\rm DT}$ = 30 ns. The APP has a maximum value of 20.3% at $V_{\rm EX}$ = 9.9 V, $V_{\rm REF}$ = 2.8 V, and $T_{\rm DT}$ = 8 ns. For $V_{\rm EX}$ = 9.9 V, the minimum APP of 2.1% is obtained at $V_{\rm REF}$ = 3.2 V and $T_{\rm DT}$ = 30 ns [9]. These DCR and APP values are smaller than those published in [4], due to the smaller area of the SPAD. The expected decrease in APP for larger V_{REF} values, i.e. for a lower detection threshold, due to a shorter quenching time and therefore a smaller avalanche charge, is verified for $V_{EX} = 9.9$ V and $T_{DT} = 8$ ns by APP = 20% for $V_{REF} = 2.8$ V and 13.4% for $V_{REF} = 3.2$ V. This is an improvement (reduction) of the APP by a factor of 1.5, which is just due to the use of a low detection threshold of 0.1 V without an increase in the dead time, i.e. without a negative impact on the possible count rate.

The PDP was characterized using a monochromator in the spectral range from 400 to 900 nm using a step size of 1 nm (see figure 3.16). With $V_{EX} = 9.9$ V, the PDP is at its maximum of 53.1% at 657 nm and a record PDP of 28.6% is achieved at 850 nm [9].

The minimum dead time of the TVQC of 7.86 ns is visible as the shortest time after which another output pulse occurs (see figure 3.17). This dead time corresponds to a maximum possible count rate of 127 Mcps [9].

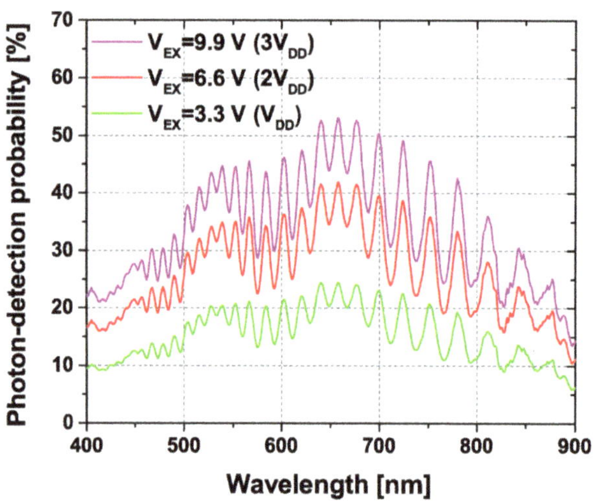

Figure 3.16. Measured PDP spectrum obtained with a triple-voltage quenching circuit using 5 V transistors for excess bias voltages of up to 9.9 V. © 2021 IEEE. Reprinted, with permission, from [9].

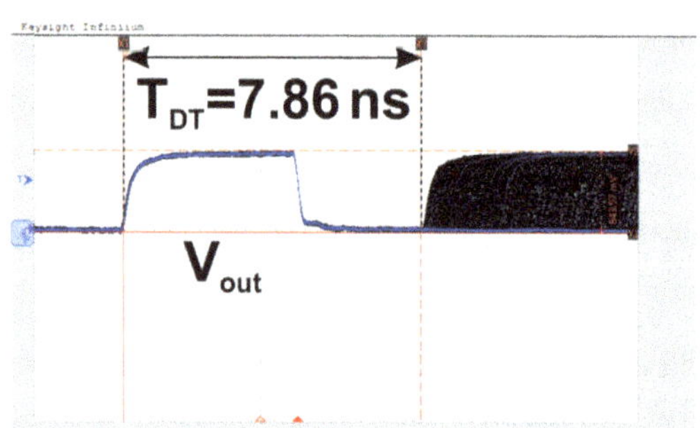

Figure 3.17. Measured output voltage transients obtained with a triple-voltage quenching circuit using 5 V transistors for excess bias voltages of up to 9.9 V, showing the dead time (no picoprobe connected). © 2021 IEEE. Reprinted, with permission, from [9].

In summary, faster quenching can be achieved with slower 5 V transistors, because fewer 5 V transistors than 3.3 V transistors are sufficient to deal with a 9.9 V excess bias. Rather low DCR and APP values were obtained for such a high excess bias voltage, due to the small SPAD area, and the APP was reduced further by early detection of avalanche events.

3.1.4 Quadruple-supply-voltage quenching circuit

The triple-voltage quenching principle using 3.3 V transistors shown in figure 3.6 was exploited with 5 V transistors to increase the excess voltage up to 13.2 V, which quadruples the nominal supply voltage of the 0.35 μm CMOS process used [12]. An excess bias of 15 V unfortunately cannot be achieved due to some voltage spikes during switching. However, according to circuit simulations, 13.2 V is possible without violating the maximum allowed specified voltages across the transistor terminals, even during switching. The circuit of the quadruple-voltage active quenching (actually passive and active (mixed) quenching) and active-resetting circuit is depicted in figure 3.18. Block A contains the SPAD with an active diameter of 40 μm, a pico-pad in a high metal layer, a double-cascoded quenching switch (MQ1–MQ4), a double-cascoded resetting switch (MR1–MR4), and SPAD biasing (MA1, MA2, whereby MA2 acts as passive quenching resistor in the early stage of avalanche events and allows sensing of the avalanche due to the voltage drop caused by the avalanche current). Block B contains the comparator with two differential amplifier stages M1–M4, an inverter as a third amplifier stage (M7 and M8) and M10, which drives the level shifter now integrated into the comparator. The level shifter after the output of the comparator of figure 3.6 was eliminated by transistors M10 and MC9 to MC12, which form a combination of cascode

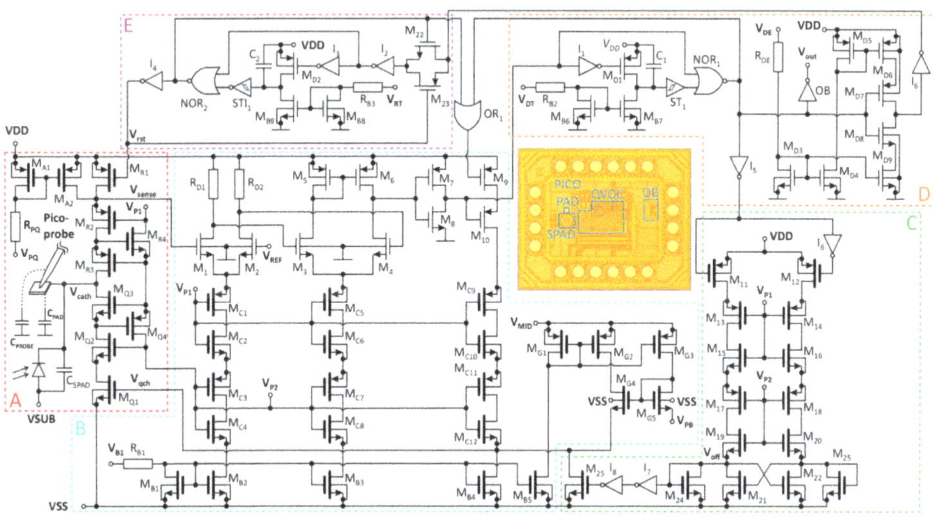

Figure 3.18. Circuit diagram of the quadruple-voltage quenching circuit using 5 V transistors for excess bias voltages of up to 13.2 V. The chip photo is inserted in the middle. © 2021 IEEE. Reprinted, with permission, from [12].

transistors, and MB4 in figure 3.18. This reduced the quenching delay, compared to that of the circuit shown in figure 3.6.

Prebiasing of the quenching transistor MQ1 is implemented (again) using transistors MG1 to MG5 (in block B) to obtain a low quenching delay. The cascoded switching stage was dimensioned for SPAD capacitances between 40 and 250 fF. With three 5 V transistors (represented by bold transistor symbols in figure 3.18) instead of six 3.3 V transistors in both the quenching and resetting switches, a lot of parasitic capacitances are avoided and the slew rate of the switches is greatly increased. In the measured fast quenching transient V_{cath} between t_3 and t_4 (see figure 3.19), the slew rate is 13.8 GV s^{-1} [12]. The transient of the cathode voltage was measured using a picoprobe placed onto the pico-pad visible in the chip photo included in figure 3.18.

The timing of the quenching and resetting circuit is controlled by the comparator's disabling circuit (block C), the quenching pulse definition circuit (block D, which also includes the output buffer), and the resetting pulse definition circuit (block E). The active area of the quadruple-voltage quenching and resetting circuit was $290 \times 210\,\mu m^2$. The mean power dissipation of the quadruple-voltage quenching and resetting chip was 16.2 mW at a dead time of 9 ns, which allows a count rate of up to 111 Mcps, reducing to 15.2 mW for a dead time of 27 ns, corresponding to a maximum count rate of 37 Mcps. In the waiting mode, the chip dissipates 11.4 mW and 10.7 mW, respectively [12].

With the picoprobe positioned on the $20 \times 20\,\mu m^2$ pad, the reaction time t_3-t_2 = 1.03 ns, the total quenching time t_4-t_2 = 2.1 ns and the resetting time t_6-t_5 = 2.8 ns were measured. It should be noticed that without the picoprobe (and without the pico-pad) the capacitance at the cathode node is lower and the reaction time is

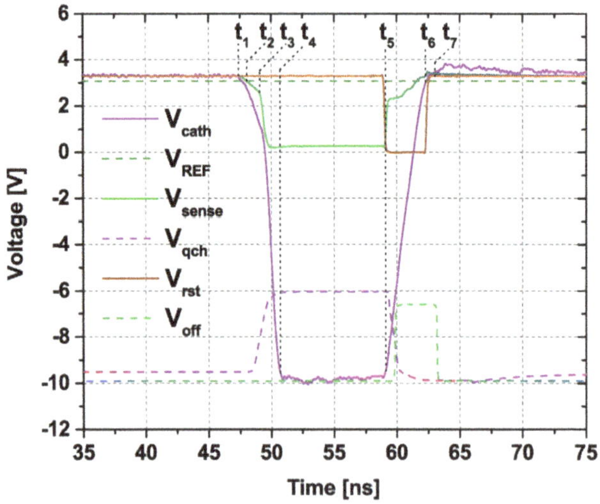

Figure 3.19. Operational principle of the quadruple-voltage quenching circuit for a dead time of 15 ns and a reference voltage of 3.15 V. The curve of V_{cath} was measured. The other curves were obtained by postlayout simulation. © 2021 IEEE. Reprinted, with permission, from [12].

0.99 ns, the quenching time is 1.9 ns and the resetting time is 2.3 ns, according to postlayout circuit simulations [12].

The DCR and APP results are shown in figure 3.20 for different reference voltages and dead times T_{DT}. Be aware that a larger reference voltage V_{ref} corresponds to a lower avalanche detection threshold. V_{ref} = 3.15 V means a detection threshold of 0.15 V (V_{DD} = 3.3 V). The maximum APP value is 36.9% for an excess voltage of 13.2 V at V_{ref} = 2.8 V (with a 0.5 V detection threshold) and a 9 ns dead time. The minimum value of the APP is 3.2% for an excess voltage of 13.2 V at V_{ref} = 3.15 V (with a 0.15 V detection threshold) and a 27 ns dead time [12]. Reducing the detection threshold from 0.5 V to 0.15 V at the maximum excess voltage for a 9 ns dead time reduces the APP from 36.9% to 27.1%.

The measured PDP spectra for four excess voltages are shown in figure 3.21. Due to the small spectral width of the light and the step size of 1 nm, the interference maximum and minima in the isolation and passivation stack above the SPAD are finely resolved and pronounced. The maximum PDP value of 67.6% is located at 652 nm and obtained for a 13.2 V excess voltage. The advantage of the thick PIN-photodiode CMOS SPAD is underlined by the PDP of 34.7% at 854 nm for the same excess voltage [12]. These are very high PDP values when we also consider the maximum possible count rate of 116 Mcps for the observed minimum dead time of 8.59 ns (see figure 3.22).

In conclusion, the low detection threshold and the low reaction time in combination with the very high slew rate of the quenching switch allow a reduction of the afterpulsing probability at very high photon detection probabilities, i.e. at very high excess voltages. The double-cascoding approach with 5 V transistors is very efficient with respect to slew rate and quenching speed. Therefore, a cheap standard 3.3 V CMOS process using 5 V transistors is sufficient, and more expensive

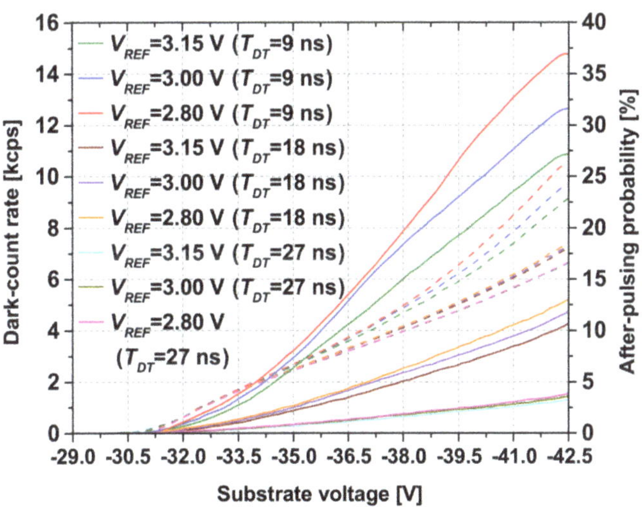

Figure 3.20. Measured DCR (dashed lines) and APP (solid lines) values obtained using the quadruple-voltage quenching circuit. © 2021 IEEE. Reprinted, with permission, from [12].

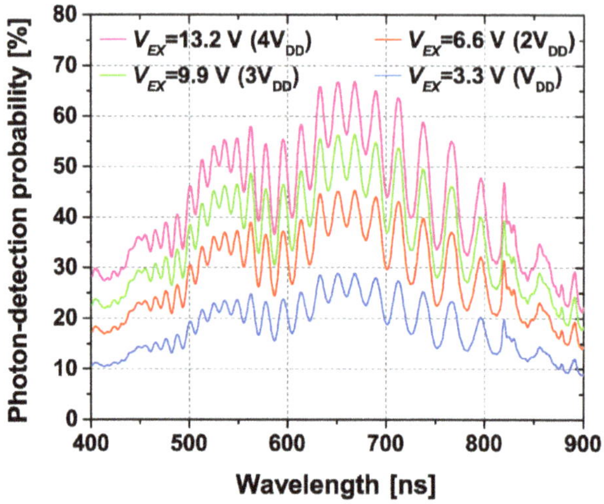

Figure 3.21. Measured PDP spectrum for the quadruple-voltage quenching circuit (reference voltage = 3.15 V and dead time = 22 ns). © 2021 IEEE. Reprinted, with permission, from [12].

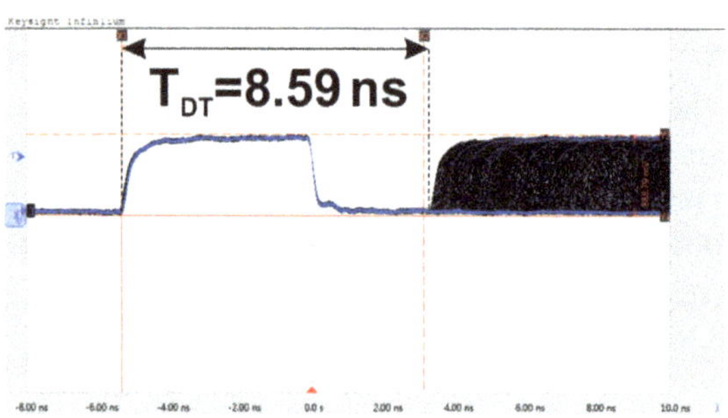

Figure 3.22. Measured minimum dead time of the quadruple-voltage quenching circuit. © 2021 IEEE. Reprinted, with permission, from [12].

high-voltage CMOS processes can be avoided for the realization of highly efficient and fast single-photon sensors and SPAD receivers.

3.2 Gating

3.2.1 Gating circuit

The principle of gating is shown in figure 3.23. When the left switch is closed, the SPAD is charged to $V_{BR} + V_{EX}$. This switch is then opened and the SPAD waits for photon absorption. If a photon is absorbed in this active gate period, the avalanche discharges the floating SPAD. A readout circuit (not shown in figure 3.23) can detect the avalanche. After the active period, the right switch in this figure is closed and the SPAD is completely discharged (if not already discharged by the avalanche current). After this inactive period, the procedure repeats. With gating, the SPAD quenches itself when fired during the active phase, or it is quenched after the active phase.

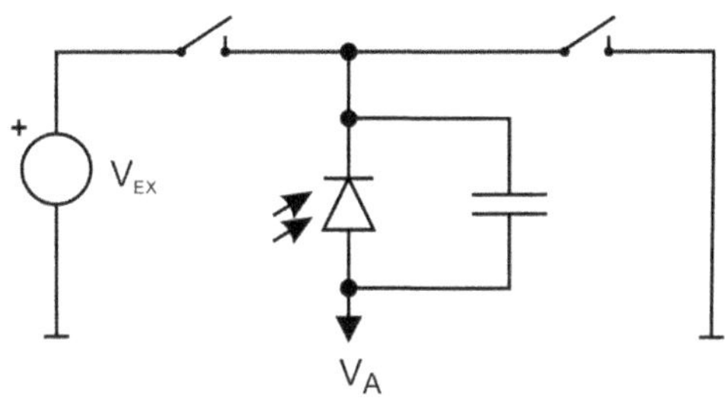

Figure 3.23. Schematic of a gating circuit.

Needless to say, gating implements active recharge. In the simplest case, each switch can be realized by one MOSFET. Gating enables high fill factors.

Gating was exploited, for instance, in [13, 14] and [15]. The pixel pitch was 24μm in a 0.35 μm high-voltage CMOS technology and a temporal resolution of 250 ps was achieved in [16]. In [14], where a 0.35μm high-voltage CMOS technology and thin SPADs were also used, a pixel pitch of 15 μm at a fill factor of 21% was reported. A 32 × 32 pixel image sensor chip for quantum physics applications was realized in 0.15 μm CMOS technology [15]. The pixel pitch was 44.64 μm at a fill factor of 19.48%. The gate windows were 50 ns at a rate of up to 800 kHz.

3.2.2 Advanced gating circuit

Because the PDP of SPADs increases with the excess voltage, it is interesting to construct switches with a higher voltage blocking capability. Cascoding can be used to obtain switches for excess voltages twice as large as the nominal supply voltage of a chip fabrication process. Figure 3.24 shows a gating circuit that exploits cascoding. A similar cascoded switch was exploited in an active quenching circuit in [17].

N0 and P0 are the switching transistors, and N1 and P1 are the cascode transistors. A negative charging pulse (from V_{DD} = 3.3 V to ground) switches on P0. P1, whose gate is grounded, then sees a large negative gate–source voltage (almost −3.3 V) and it also switches on. The cathode of the SPAD is connected to V_{DD}, and its cathode is charged to $V_{CAT} = V_{DD}$. The SPAD is made active, and the active gate phase starts. P0 can now be switched off. N0 is off during charging and during the active phase. N1 prevents the drain–source voltage of N0 from becoming larger than its breakdown voltage during the active phase. Because the gate of N1 is grounded, its source potential is also at ground (because no current flows through N0). Therefore, the voltage of $V_{DD} - V_{SS} = 3.3V - (-3.3V) = 6.6V$ is equally divided between N0 and N1.

At the end of the active gate phase, a positive discharging pulse (from −3.3 V to ground) switches on N0. N1 now sees a positive gate–source voltage of almost 3.3 V and also switches on. If no photon has triggered an avalanche or a triggered avalanche has not yet discharged the SPAD during the active phase, N0 and N1 now discharge the SPAD and the non-active phase starts. Analogously, the grounded gate of P1 limits the drain–source voltage of P0 to −3.3 V and the drain–source voltage of P1 is also −3.3 V during the inactive phase of the SPAD.

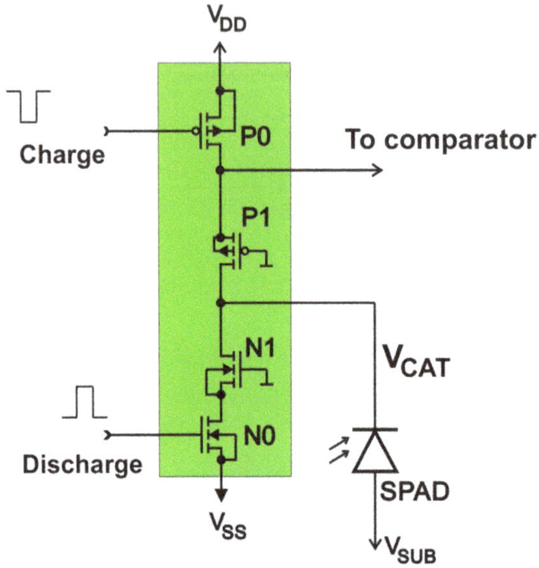

Figure 3.24. Schematic of an advanced gating circuit that applies the cascoding technique.

In this way, the cascoded switch allows a SPAD excess voltage of up to twice the nominal supply voltage of a CMOS technology (6.6 V in 0.35 μm CMOS with a nominal supply voltage of 3.3 V). Of course, a triple-well CMOS technology is necessary to enable the connections of the bulks (wells) of NMOS and PMOS transistors to their sources. The grounded gate of P1 also protects the comparator from an overvoltage at its input.

A comparator is necessary to detect whether a photon has triggered an avalanche during the active phase. The comparator usually decides this after the end of the active phase. A low decision threshold is advantageous in order to also detect avalanches which were triggered close to the end of the active phase and which therefore did not have enough time to grow much (see figure 1.36, 1.37 and 1.42).

Such an advanced cascoded gater was used to characterize SPADs with respect to avalanche build-up [18, 19]. Some results of this characterization are shown in figures 1.36 to 1.39 and in figures 1.37–1.46. An advanced cascoded gater was also implemented in a SPAD receiver [20]. As a result of doubling the nominal supply voltage of 3.3 V to 6.6 V, excess voltages of up to 6.6 V were possible, two detected photons were sufficient for a ‘1’ bit, and a 12.7 dB gap to the quantum limit was achieved with a cascoded gater. The receiver circuit and the results obtained are described in figures 4.20–4.23.

References

[1] Dervić A, Goll B, Steindl B and Zimmermann H 2019 Transient measurements and mixed-quenching, active-resetting circuit for SPAD in 0.35 μm high-voltage CMOS for achieving 218 Mcps *2019 26th IEEE Int. Conf. on Electronics, Circuits and Systems (ICECS)* pp 819–22

[2] Steindl B, Hofbauer M, Schneider-Hornstein K, Brandl P and Zimmermann H 2018 Single-photon avalanche photodiode based fiber optical receiver up to 200 Mb/s *J. Sel. Topics Quantum Electron.* **24** 3801308

[3] Enne R, Steindl B, Hofbauer M and Zimmermann H 2018 Fast cascoded quenching circuit for decreasing afterpulsing effects in 0.35 μm CMOS *IEEE Solid-State Circ. Lett.* **1** 62–5

[4] Dervic A, Steindl B, Hofbauer M and Zimmermann H 2019 High-voltage active quenching and resetting circuit for SPADs in 0.35μm CMOS for raising the photon detection probability *Opt. Eng.* **58** 40501–4

[5] Webster E A G, Grant L A and Henderson R K 2012 A high-performance single-photon avalanche diode in 130-nm cmos imaging technology *IEEE Electron Device Lett.* **33** 1589–91

[6] Niclass C *et al* 2007 A 130-nm CMOS single photon avalanche diode *Proc. SPIE* **6766** 676606

[7] Villa F *et al* 2014 CMOS SPADs with up to 500μm diameter and 55% detection efficiency at 420 nm *J. Mod. Opt.* **61** 102–15

[8] Ceccarelli F *et al* 2018 152-dB dynamic range with a large-area custom-technology single-photon avalanche photodiode *IEEE Photonics Technol. Lett.* **30** 391–4

[9] Dervić A, Hofbauer M, Goll B and Zimmermann H 2021 Integrated fast-sensing triple-voltage SPAD quenching/resetting circuit for increasing PDP *IEEE Photonics Technol. Lett.* **33** 139–42

[10] Ceccarelli F *et al* 2019 Fully integrated active quenching circuit driving custom-technology SPADs with 6.2-ns dead time. *IEEE Photonics Technol. Lett.* **31** 102–5

[11] Acconcia G *et al* 2016 High-voltage active quenching circuit for single photon count rate up to 80 Mcounts/s *Opt. Express* **24** 17819–31

[12] Dervić A, Hofbauer M, Goll B and Zimmermann H 2021 High slew-rate quadruple-voltage mixed-quenching active-resetting circuit for SPAD in 0.35-μm CMOS for increasing PDP *IEEE Solid-State Circ. Lett.* **4** 18–21

[13] Maruyama Y, Blacksberg J R and Charbon E 2013 A 1024×8 700 ps time-gated spad line sensor for laser raman spectroscopy and libs in space and rover-based planetary exploration *2013 IEEE International Solid-State Circuits Conf.* pp 110–1

[14] Perenzoni M, Nicola Massari, Perenzoni D, Gasparini L and Stoppa D 2016 A 160×120 pixel analog-counting single-photon imager with time-gating and self-referenced column-parallel a/d conversion for fluorescence lifetime imagingg *IEEE J. Solid-State Circ.* **51** 155–67

[15] Gasparini L, Zarghami M, Xu H, Parmesan L, Garcia M M, Unternährer M, Bessire B, Stefanov A, Stoppa D and Perenzoni M 2018 A 32×32-pixel time-resolved single-photon image sensor with 44.64μm pitch and 19.48% fill-factor with on-chip row/frame skipping features reaching 800 kHz observation rate for quantum physics applications *IEEE Int. Solid-State Circuits Conf.* pp 98–9

[16] Maruyama Y, Blacksberg J and Charbon E 2014 A 1024×8 700-ps time-gated SPAD line sensor for planetary surface exploration with laser Raman spectroscopy and LIBS *IEEE J. Solid-State Circ.* **49** 179–89

[17] Zimmermann H, Steindl B, Hofbauer M and Enne R 2017 Integrated fiber optical receiver reducing the gap to the quantum limit *Sci. Rep.* **7** 2652

[18] Goll B, Hofbauer M, Steindl B and Zimmermann H 2018 Transient response of a 0.35μm CMOS SPAD with thick absorption zone *2018 25th IEEE Int. Conf. on Electronics, Circuits and Systems (ICECS)* pp 9–12

[19] Goll B, Steindl B and Zimmermann H 2020 Avalanche transients of thick 0.35μm CMOS single-photon avalanche diodes *MDPI Micromachines* **11** 869 (Special Issue “Miniaturized Silicon Photodetectors: New Perspectives and Applications”)

[20] Goll B, Hofbauer M, Steindl B and Zimmermann H 2018 A fully integrated SPAD-based CMOS data-receiver with a sensitivity of -64 dBm at 20 Mb/s *IEEE Solid-State Circ. Lett.* **1** 2–5

IOP Publishing

Single-photon Detection for Data Communication and Quantum Systems

Michael Hofbauer, Kerstin Schneider-Hornstein and Horst Zimmermann

Chapter 4

SPAD receivers for data communications

In this chapter, we first introduce the quantum limit that follows from the Poisson statistics. We then present a statistical investigation of single-photon avalanche diode (SPAD) parasitic properties and the results of a bit error ratio model for SPAD receivers. Thereafter, two few-channel SPAD receivers in p–intrinsic–n (PIN)-photodiode CMOS technology are explained. We compare analog and digital processing of the quencher output signals. In addition, a few-channel SPAD receiver that uses high-voltage (HV) CMOS is described, in addition to a gated SPAD receiver with only one SPAD that exploits a sub-bit photon-counting principle. It is shown that two photon detections within one bit for a logical '1' are sufficient to achieve a bit error ratio that is lower than the error-correction limit. After a comparison of the sensitivities of SPAD and APD receivers and the remaining distances to the quantum limit, we present optical wireless communications experiments with SPAD receivers and their results.

4.1 Modeling of receiver bit error ratio

The fundamental physical limit for the sensitivity (the optical power necessary to achieve a certain bit error rate (BER)) of optical receivers is given by the Poisson statistics. This fundamental limit is also called the quantum limit. This quantum limit restricts optical receivers to sensitivities that are about 25–30 dB better than PIN-FET (PIN-photodiode field-effect transistor) receivers [1] and about 15–20 dB better than APD receivers (receivers with avalanche photodiodes (APDs) in the linear mode) [2]. Single-photon avalanche diodes (SPADs) possess very high gains (exceeding 10^6). Due to the very high gain of SPADs, the electronic noise caused by amplifiers (shot noise, thermal noise) and the excess noise of APDs are no longer limiting factors. A SPAD can readily be connected to a digital gate. Therefore, SPAD receivers represent digital optical receivers. In principle, they do not need

doi:10.1088/978-0-7503-2584-4ch4

(analog) amplifiers, and, in this respect, they are unlike receivers with PIN photodiodes or APDs.

We will examine the Poisson statistics first. According to the Poisson statistics, the number of photons varies a little within the duration $T = 1/B$, where B is the bit rate per second. The probability $p_m(k)$ that k photons will hit the detector in T, such that m photons arrive on average (this determines the average optical power $\langle P_{\text{opt}} \rangle$ and the sensitivity) in T, can be calculated from the Poisson distribution [3]:

$$p_m(k) = \frac{m^k}{k!}e^{-m}. \tag{4.1}$$

For an ideal detection efficiency of 100%, on average, 21 photons ($m = 21$) are necessary in a '1' bit to achieve a BER = 10^{-9}, because the probability that a '0' ($k = 0$) is detected has to be less than 10^{-9}. Error correction nowadays allows BER = 2×10^{-3}, meaning that only seven photons are necessary, on average. However, we have to be aware that an optical transmitter with an extinction ratio of infinity is necessary, i.e. for a '0,' no photons must be emitted. Therefore, SPAD receivers used as receivers in optical wireless communications suffer strongly from the effects of ambient light.

The average optical power is given by ($h\nu$ is the photon energy):

$$<P_{\text{opt}}> = \frac{mh\nu B}{2}. \tag{4.2}$$

The sensitivity S in dBm is represented by:

$$S = 10\log\left(\frac{<P_{\text{opt}}>}{1\ \text{mW}}\right), \tag{4.3}$$

however, we must also mention the BER for which S is given. Figure 4.1 compares the sensitivities corresponding to the quantum limit for BER = 10^{-9} and BER = 2×10^{-3}. There is a difference of about 5 dB between the quantum limits for these two

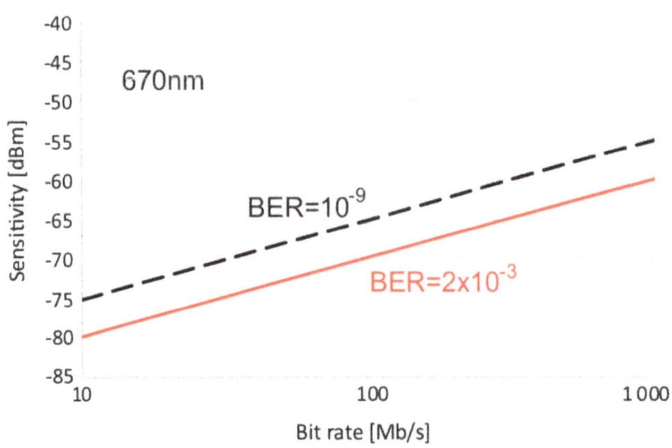

Figure 4.1. Quantum limit for λ=670 nm.

BERs. It should be noted that the quantum limit shifts with the wavelength λ (because of ν in equation (4.2), $\lambda = c/\nu$, c = light velocity).

However, SPADs are not ideal devices. They are subject to dark counts, afterpulsing, and optical crosstalk. (During an avalanche event, hot carriers cause the emission of photons from the multiplication zone of a fired SPAD. These photons can be absorbed by neighbouring SPADs and can trigger an avalanche there, if the SPADs are arranged in arrays.) These parasitic effects increase the BER of SPAD receivers. In order to describe the influence of these parasitic effects, a model for the BER of SPAD receivers was developed [4]. The formulation of the equations required is quite lengthy, and the interested reader is referred to the original publication. This model is quite general, i.e. it is independent of the number of SPADs in a detector array. It considers the dark count (DC) rate, the afterpulsing (AP) probability, and the optical crosstalk (OCT) probability, including their combinations, during bit reception.

When the avalanche events in a four-SPAD receiver were statistically investigated (measured and stored using a four-channel oscilloscope, see [5]), the DCs, APs, and OCTs could be easily identified [6]. Figure 4.2 shows the interarrival time distribution of one of the SPADs in the four-SPAD receiver described in [5]. Afterpulsing and dark counts can be clearly distinguished, because the DCR of the order of 10^4 s^{-1} leads to a medium interarrival time (the time between two subsequent events) of 10^5 ns (although there is a wide spread from 10^3 ns to almost 10^6 ns between two dark counts) and because the exponential decay time constant of afterpulsing is on the order of 10 ns. A random-to-arrival (the time interval between a randomly chosen instant t_0 and the first arrival after t_0) evaluation identified only the distribution due to dark counts [6].

Figure 4.3 shows the detection probabilities of the statistical evaluation of avalanche events in two neighbouring SPADs (SPAD1 and SPAD2) in the array of four SPADs in the four-SPAD receiver reported in [5]. For a random selection of t_0 and a nanosecond interval Δt (2 ns), the detection probability of dark counts is $\Delta t/\tau_{dc}$

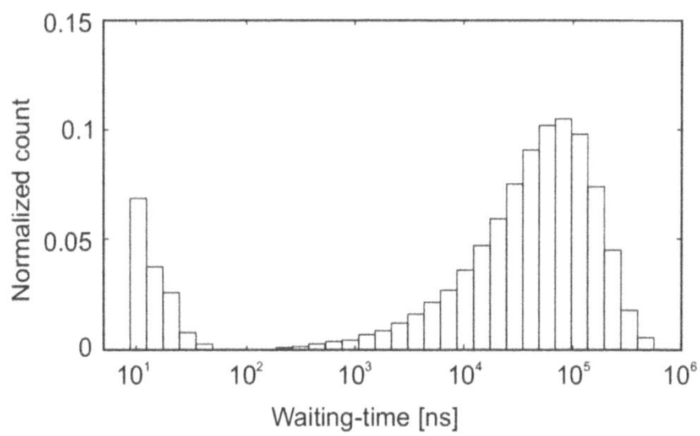

Figure 4.2. Interarrival time distribution between avalanche events in darkness for one of the SPADs of the four-SPAD receiver presented in [6].

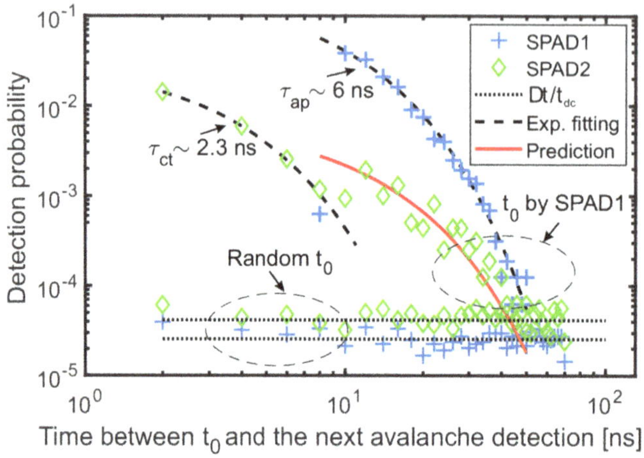

Figure 4.3. Waiting-time distribution for interarrival and random-to-arrival evaluations for two neighbouring SPADs in the array of four SPADs. © 2019 IEEE. Reprinted, with permission, from [6].

(τ_{dc} = $1/r_{dc}$ with a dark count rate of r_{dc}), leading to the two horizontal lines for SPAD1 and SPAD2 in figure 4.3 [6].

Setting t_0 to detection instants in SPAD1 (whereby only instants were allowed for which no avalanche event occurred in all four SPADs during the preceding 100 ns, in order to exclude afterpulses and crosstalk), increases the detection probabilities considerably, compared to a random choice of t_0. Crosstalk into SPAD2 becomes visible during the quencher dead time of SPAD1, which is 9 ns. A good match between these crosstalk events in SPAD2 and an exponential decay with a time constant of τ_{ct}=2.3 ns can also be seen in figure 4.3. After the dead time of 9 ns, afterpulsing can occur in SPAD1, and these events obey an exponential distribution with a time constant of τ_{ap}=6 ns. The third, solid red curve corresponds to afterpulses in SPAD2 that occur after crosstalk avalanches in SPAD2.

The crosstalk was investigated in more detail [6]. The average one-to-one crosstalk probabilities, determined as described above, and the waiting time, i.e. delay, are plotted in the left part of figure 4.4. The crosstalk increases with the excess bias voltage, as expected. The crosstalk delay, however, decreases for lower excess bias voltages and increases slightly above 4.5 V.

In a SPAD receiver, crosstalk events other than those from one SPAD to another SPAD can happen. Therefore, the probabilities of firing i SPADs through crosstalk when j SPADs are already fired by another mechanism were also investigated [6]. These conditional crosstalk probabilities $P(i|j)$ were calculated by assuming that crosstalk between SPADs is a Bernoulli process, i.e. the probability of triggering an avalanche in any available SPAD is p_{cr} and the probability that no avalanche is triggered is (1-p_{cr}) [7, 8]. For instance, in an array of four SPADs, $P(0|1) = (1\text{-}p_{cr})^3$ is the conditional probability that i = 0 crosstalks are triggered when j = 1 SPAD is initially fired. In addition, cascaded processes were included in the crosstalk model, i.e. a crosstalk-triggered avalanche can trigger a following avalanche

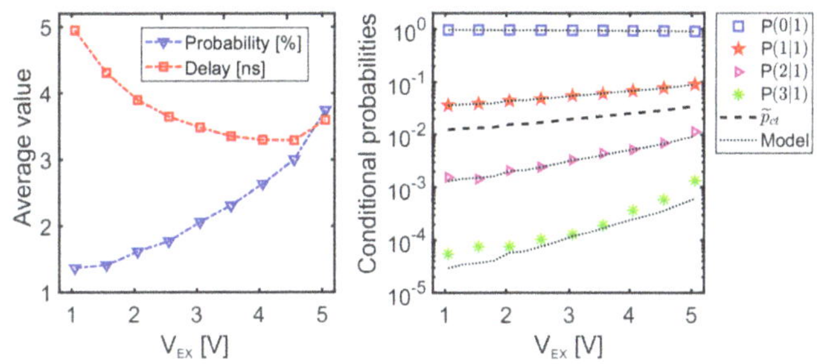

Figure 4.4. SPAD-to-SPAD crosstalk probability and delay in the array of four SPADs (left) and a comparison of modelled and measured conditional crosstalk (right), both in dependence on the excess bias voltage. © 2019 IEEE. Reprinted, with permission, from [6].

in another SPAD that is still available in the array. As an example, $P(1|1) = 3p_{cr}(1 - p_{cr})^2(1 - p_{cr})^2$ is the probability that one crosstalk (i = 1) will occur when one SPAD (j = 1) is initially triggered. (Three is the number of available SPADs, each with the probability p_{cr}, among which, two are not triggered, leading to the first of the two terms $(1\text{-}p_{cr})^2$. The second of these two terms comes from the cascaded process, in which none of the two still-available SPADs is fired by the third SPAD, which underwent the crosstalk.) The conditional probabilities $P(2|1)$, $P(3|1)$, $P(0|2)$, $P(1|2)$, $P(2|2)$, $P(0|3)$, and $P(1|3)$ also have to be considered for an array of four SPADs [6]. Therefore, modelling the crosstalk correctly is quite complex, especially if an array contains more than four SPADs.

This crosstalk model was verified by a comparison with the experimental dark noise data. The $P(i|j = 1)$ values obtained were averaged over all SPADs as initializing (triggering) SPADs for i = 0, 1, 2, and 3, giving the dotted curves in figure 4.4(b). The period of 8 ns after each avalanche event was considered to catch the cascaded events. There is a very good agreement between the measured conditional crosstalk probabilities and the model calculations. This crosstalk model was implemented in the complete BER model [4].

As part of the BER model, the probability $P_z('1')$ that no photon is counted in a '1' bit was compared to a Monte Carlo (MC) experiment (see figure 4.5). Here, 10^4 equally likely logical '0' and '1' bits were generated and the photon arrivals were determined by random numbers. The dead time of the SPAD was considered. The results in figure 4.5 show that the developed model describes the photon-count statistics well and is appropriate for modeling the BER of SPAD receivers. It should also be mentioned that there is a minimum for $P_z('1')$ when λ_1 T_1 equals about 6.

For the BER model, the optical power, bit rate, excess bias voltage, and the dependence of the DCR, APP, and crosstalk on the excess bias voltage were needed and were available from measurements. In addition, the photon detection probability PDP of the SPAD(s) must be available. In fact, the optical fill factor of the SPAD array can be included if the PDE (photon detection efficiency) is used instead of the PDP. The dependence of the PDE on the excess bias voltage can be estimated by $\text{PDE}(V_{ex}) = \text{PDE}_0(1 - e^{V_{ex}/V_0})$ [9], where PDE_0 is the saturation value for large excess bias voltages and V_0 is a normalization coefficient. These parameters were

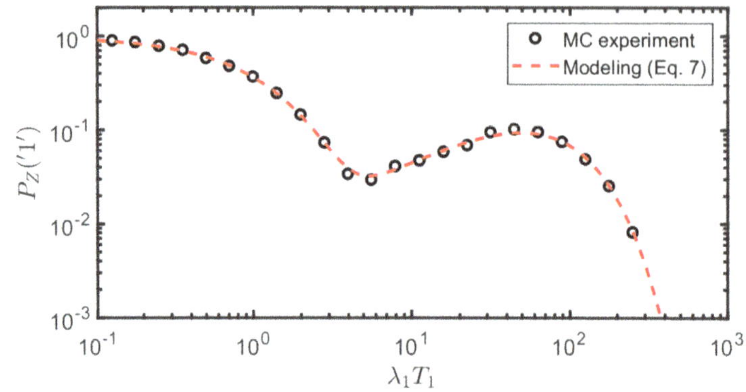

Figure 4.5. Error probability of a logical '1' for a dead time of 40% of the bit duration T_b, a return to zero (RZ) duty cycle T_1/T_b of 20% and an extinction ratio of 200 (λ_1 is the photon-count rate per second in a logical '1' and T_1 is the duration of a light pulse). Copyright 2018 IEEE. Reprinted, with permission, from [4].

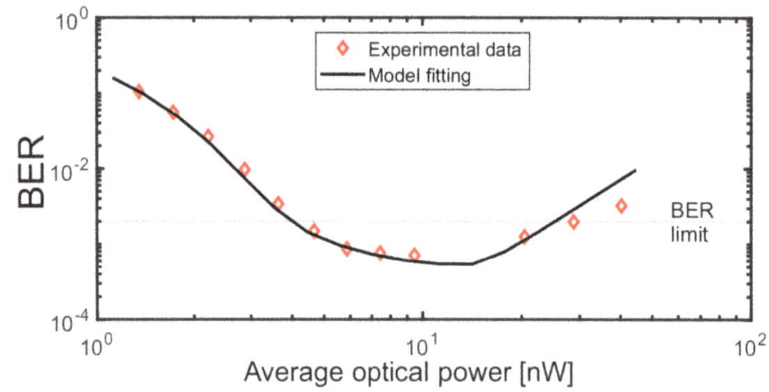

Figure 4.6. Comparison of modelled and measured BERs for 50 Mb s^{-1} with the four-SPAD receiver presented in [4, 5]. Copyright 2018 IEEE. Reprinted, with permission, from [4].

fitted in the BER model. Figure 4.6 compares the modeled BER to the measured values with a detection threshold of four photon detections for a '1,' a duty cycle of 60%, a dead time of 9 ns, and a light-source extinction ratio of 200. As in the experiment, the excess bias voltage was varied for each optical power to obtain the minimum BER. There is excellent agreement between the modeled and measured BERs. Therefore, the BER model was used to investigate the influences of the different noise effects, such as the DCR, APP, and OCT. In fact, this was the purpose of developing the BER model for the SPAD receiver(s), i.e. to save the large amount of time needed for extensive experiments and to perform investigations that are not possible in experiments.

A sensitivity analysis was performed using the BER model by changing one factor at a time (OFAT) and calculating the change to the minimum possible BER. The modeling results are presented in figure 4.7. Afterpulsing has a much stronger effect on the BER than dark counts, but the most important noise effect originates from crosstalk, which is much larger than that from afterpulsing. A further result of [4]

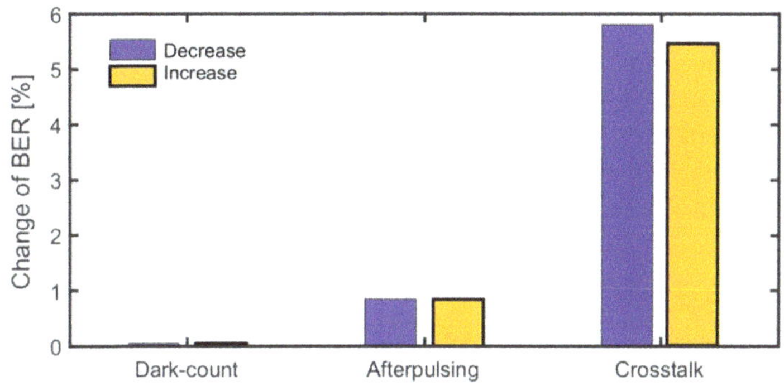

Figure 4.7. Sensitivity of the minimum possible BER at 50 Mb s^{-1} for ± 10% change in the DCR, APP, and OCT. Copyright 2018 IEEE. Reprinted, with permission, from [4].

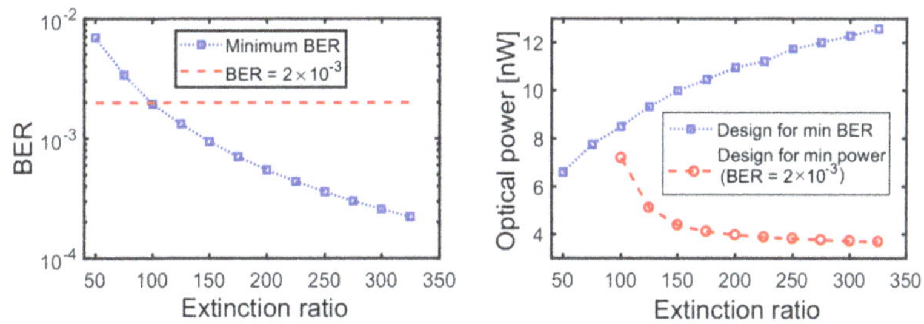

Figure 4.8. BER and average optical power at 50 Mb s^{-1} for a minimum-BER design and for a minimum-power design. Copyright 2018 IEEE. Reprinted, with permission, from [4].

was that a reduction of the one-to-one crosstalk by a factor of two would allow a reduction of the detection threshold for a logical '1' from four to three photon detections, i.e. it would lead to an improvement in the sensitivity.

Furthermore, the influence of the light source's extinction ratio on the BER was studied using the BER model. The left part of figure 4.8 shows that an extinction ratio of at least 100 is necessary for a BER of 2×10^{-3}. This is quite bad news for outdoor optical wireless communications (OWC) with SPAD receivers; even for indoor OWC, SPAD receivers will have to be equipped with optical bandpass (interference) filters.

There are two possibilities for SPAD receiver designs: optimization of the optical power to achieve the minimum BER and minimization of optical power for a given BER (which was chosen as the BER limit of 2×10^{-3}). The right part of figure 4.8 shows the dependence of the optical power on the light-source extinction ratio for these two design possibilities. About 8 nW is necessary for an extinction ratio of 100. For an increasing extinction ratio, less optical power is required by the minimum-power design. The power reduction, however, is not exponential, as would be expected from a first-order BER model based on Poissonian photon-counting statistics [4]. The reasons for this are the parasitic events (DCR, APP, and OCT), which are considered properly in the developed BER model.

In summary, the BER model considers all the important, non-ideal effects, such as the dead time, light-source extinction ratio, PDP, DCR, APP, and crosstalk as well as their dependence on the excess bias voltage. An important result of this SPAD receiver model development is that optical crosstalk between the SPADs in a four-SPAD receiver has a much larger effect on the BER than afterpulsing and dark counts [4], as confirmed by a sensitivity analysis using the developed BER model, whose results are presented in figure 4.7. In a practical application of the BER model, the difficulty of obtaining the crosstalk probabilities for multi-SPAD receivers would be much larger, and a model for the photon emission by hot carriers in an avalanche would be needed. An interesting conclusion might be that optical crosstalk in the multi-SPAD receiver presented in [10] is the reason for its moderate BER performance.

4.2 Fiber receivers

SPADs are of great interest for use in optical receivers because of their very high gain, which is more than a million in the Geiger mode. This offers the possibility of eliminating electronic noise (shot noise and the thermal noise of transistors) and the excess noise of APDs and therefore of improving the sensitivity to levels far above those of PIN-photodiode receivers and linear-mode APD receivers.

A high-dynamic-range 0.13 μm CMOS SPAD-based optical receiver with a matrix of 32 × 32 thin SPADs was suggested in [11]. With 450nm light, it achieved a BER of 10^{-9} at a data rate of 100 Mb s^{-1} with a sensitivity of −31.8 dBm.

If a high dynamic range is not the primary goal, and if it is sufficient to achieve a BER limit of 2×10^{-3} for error correction [12, 13], many fewer SPADs will suffice in a receiver. One SPAD, offering one-photon detection for a logical '1' bit, will not be sufficient at high data rates, because of the DCR and mainly because of the APP, which is usually in the percent region, leading to a BER that is above the error-correction limit. The first guess for the thick SPAD (see section 1.3.2) was that four SPADs and the detection of one photon in each of them for a logical '1' should allow a BER below the error-correction limit.

A four-SPAD receiver was designed in 0.35 μm PIN-photodiode CMOS and the results were published in [5]. A chip photo of this receiver is shown in figure 4.9. The array of four thick SPADs had an optical fill factor of 53%. The diameter of the SPAD array was 200μm. The cascoded active quenching circuit depicted in figure 3.4 was implemented for each SPAD. The outputs of the quenching circuits were transferred by four integrated output buffers to a 50 Ω output impedance connected to the inputs of a four-channel five-gigasamples-per-second (GSPS) oscilloscope. The experiments were performed with a high-extinction-ratio light source (a continuous-wave (CW) laser with an external modulator) with a wavelength of 635 nm. The received data were stored in the oscilloscope, and post-processing was done in MATLAB to determine the bit error ratio (for details, please see [5]).

The results for the dependence of the bit error ratio on the average optical input power are shown in figure 4.10. Digital and analog post-processing were compared, leading to the conclusion that analog post-processing led to better sensitivities: −54 dBm, i.e. 4 nW for BER = 2×10^{-3} at 50 Mb s^{-1} in NRZ and −55.7 dBm, i.e. 2.7 nW in

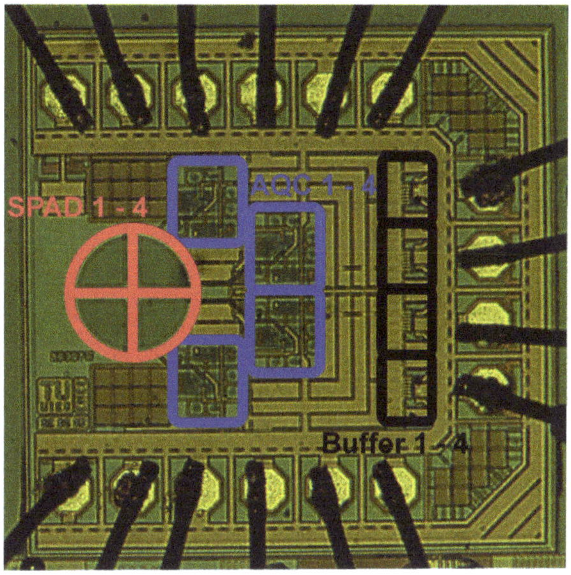

Figure 4.9. Chip photo of the four-SPAD receiver reported in [5]. (2017) Copyright. With permission of Springer.

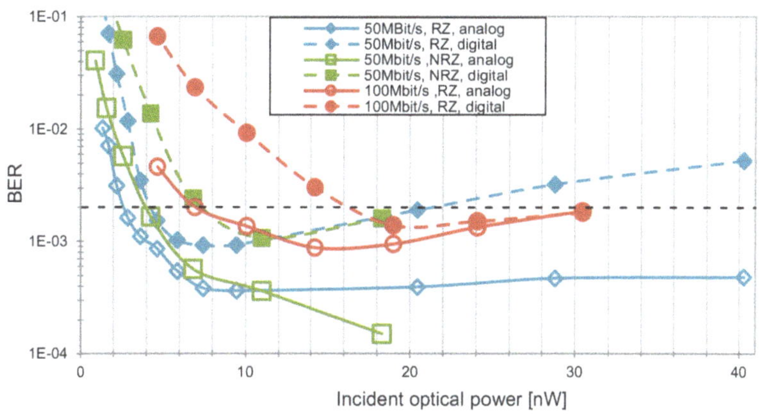

Figure 4.10. Dependence of the BER of the four-SPAD receiver on average optical input power (non-return to zero (NRZ) and return to zero (RZ)) [5].

RZ with a duty ratio of 50%. A sensitivity of −51.6 dBm, i.e. 7 nW, for BER = 2×10^{-3} was achieved at 100Mb s^{-1} in RZ with a duty ratio of 10% using analog post-processing [5]. The small duty ratio at 100Mb s^{-1} was necessary because of the dead time of 9 ns (because one bit at 100 Mb s^{-1} lasts only 10 ns). For the first time, these sensitivities were better than those of integrated linear-mode APD receivers built using the same CMOS technology [5].

Although analog post-processing showed better results, digital post-processing was easier to implement on the SPAD receiver chip, which was realized in the same 0.35 μm

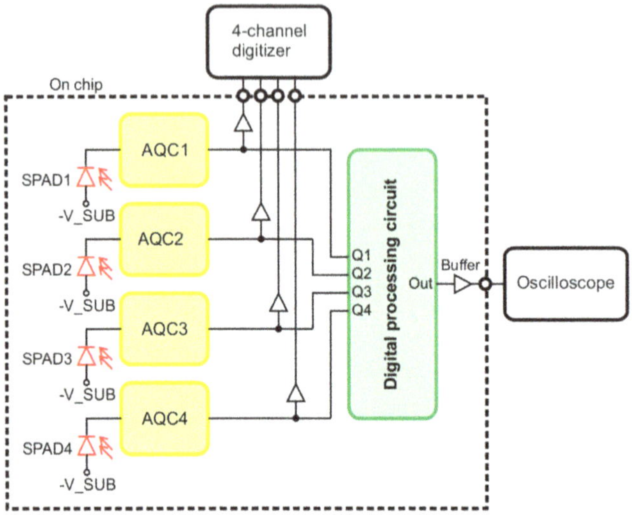

Figure 4.11. Block diagram of the four-SPAD receiver with integrated digital processing described in [14].

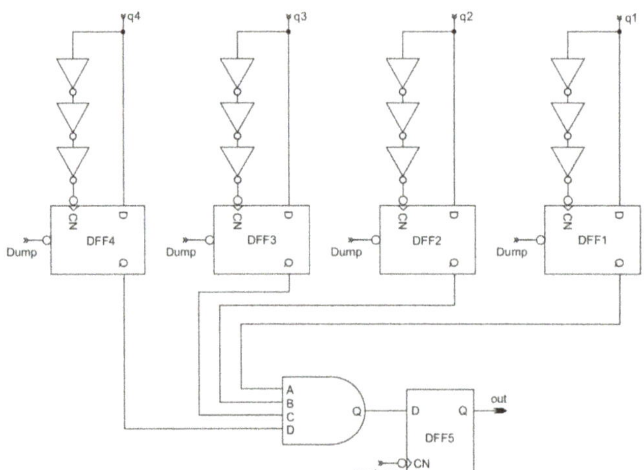

Figure 4.12. Latch-type digital processing circuit of the four-SPAD receiver presented in [14].

PIN-photodiode CMOS technology that was used to produce the thick SPAD. The block diagram of this receiver is depicted in figure 4.11 [14]. The active quenching circuits were slightly modified, compared to the cascoded quenching circuit described above (see figure 3.4). All four quencher outputs were made available via 50Ω buffers in order to enable different post-processing methods. The quencher outputs, however, were also connected to the on-chip digital processing circuit, whose output was also buffered. The diameter of the four-quadrant SPAD was reduced to 117 μm and the dead time was shortened to 3.5 ns [14].

The circuit diagram of the digital processing circuit is presented in figure 4.12. An incoming logical '1' is latched by the D-flip-flops DFF*i* (i = 1, 2, 3, 4) because the three inverters in front of the clock input CN generate a falling edge shortly after the quencher outputs switch from low to high to indicate the absorption of a photon. The 'Dump' input enables the DFFs during a bit and resets them at the end of a bit. Only if all four DFFs latch a logical one in a bit does the AND gate set DFF5. Readout is performed 1.5 ns before the latch reset to allow a long enough time for the receiver to be sensitive [14]. As a consequence, a logical '1' at the output of the receiver needs a photon to be detected by each of the four SPADs during the bit duration. This processing only allows the detection of one photon per SPAD within a bit duration, and is therefore disadvantageous if the bit duration is much longer than the dead time. However, this receiver was aiming for the highest possible data rate. It also should be mentioned that with a dead time of 3.5 ns, this quencher allows a maximum photon-count rate of almost 300 Mcps in a SPAD.

A chip photo of this four-SPAD receiver can be seen in figure 4.13. The receiver chip has dimensions of $1400 \times 1040\,\mu$m. The digital signal processing circuit only occupies an area of 0.014 mm^2 [14].

For data transmission experiments using a single-mode fiber and a 635 nm single-mode laser source, a return-to-zero signal that had a duty ratio of 20% and PRBS7 were used [14]. Figure 4.14 shows the BER as a function of the optical input power. The curves '1 SPAD' to '4 SPADs' were obtained from the four direct quencher outputs using MATLAB post-processing when one or more SPADs detected a photon during a bit. Four photon detections were necessary to obtain a BER that was less than the error-correction limit of 2×10^{-3} (represented by the dash-dotted line in figure 4.14). The sensitivity for 50Mb s^{-1} and BER=2×10^{-3} was 7.6 nW (−51.2 dBm). The sensitivity with the integrated digital post-processing was about the same (−51.4 dBm) [14]. Only the BER was slightly larger than for the MATLAB post-processing, because part of the bit duration was required for readout—whereas the MATLAB processing was ideal and used the whole bit duration.

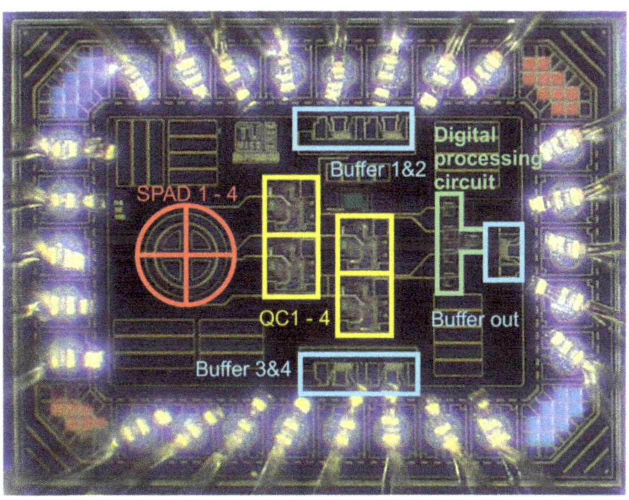

Figure 4.13. Chip photo of the four-SPAD receiver with integrated digital processing. © 2018 IEEE. Reprinted, with permission, from [14].

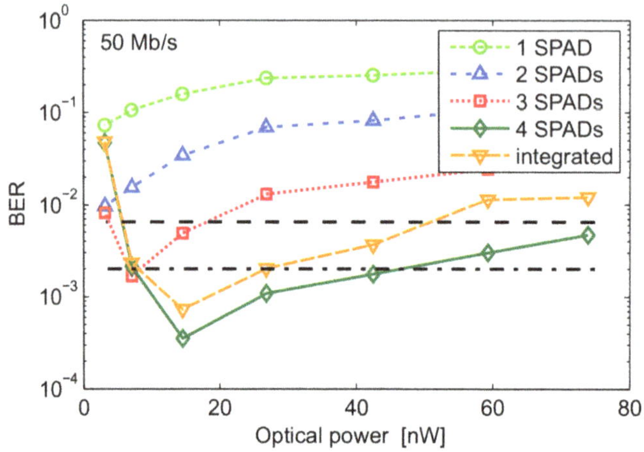

Figure 4.14. Dependence of BER on average optical power for 50 Mb s^{-1}. © 2018 IEEE. Reprinted, with permission, from [14].

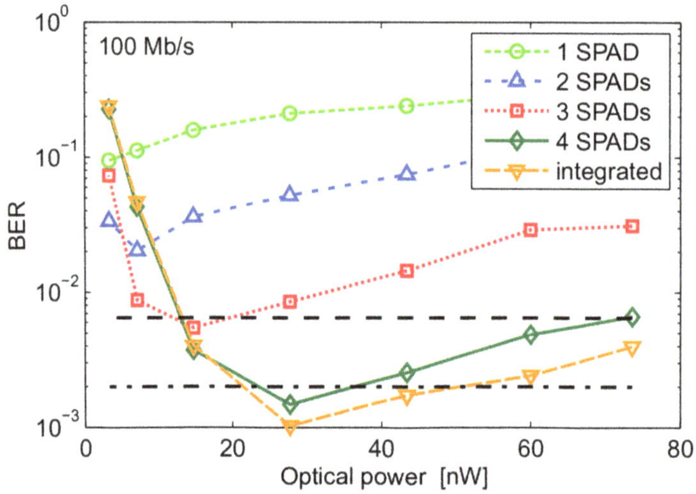

Figure 4.15. Dependence of BER on average optical power for 100 Mb s^{-1}. © 2018 IEEE. Reprinted, with permission, from [14].

At 100 Mb s^{-1} (see figure 4.15), four photon detections are also necessary to obtain a BER below the error-correction limit of 2×10^{-3}. The sensitivity for 100 Mb s^{-1} and BER = 2×10^{-3} is 23.5 nW (−46.3 dBm).

For 150 and 200 Mb s^{-1}, the BER did not reach the BER limit of 2×10^{-3} with the use of concatenated Reed-Solomon and product codes [13]. However, with super forward error correction (FEC) and a BER of 6.5×10^{-3} [13], sensitivities of −46.1 dBm and -43.7 dBm were obtained for 150 and 200Mb s^{-1}, respectively, both using MATLAB post processing [14].

Due to the possibility of implementing an antireflective coating (ARC), a four-SPAD receiver was realized in 0.35 μm HV CMOS [15]. The HV CMOS technology used was part of the same modular CMOS process family as the PIN-photodiode CMOS process used for the two four-SPAD receivers described above. This enables a direct comparison of the SPAD receivers constructed using both technologies.

The HV CMOS SPAD described in chapter 1 was implemented in the HV 4-SPAD receiver IC. With an ARC, this HV CMOS SPAD possessed a PDP of approximately 45% at 635 nm and an excess bias of 6.6 V [16]. Quenching circuits with a low detection threshold were integrated, which were similar to the quenchers already described in detail [5]. Cascoding increased the maximum excess bias voltage from the usual supply voltage of 3.3–6.6 V, and the dead time was adjustable between 5.8 and 34 ns. The output stage of each of the four channels consisted of a 50Ω output driver. A chip photo of the fabricated test chip is shown in figure 4.16. The area of the chip was 1160×2120 μm^2. In the sensitive state without any detection, the average power consumption of the four quenching circuits was about 24.4 mW. Simultaneous photon detection in each of the four SPADs increased the power consumption to 40.4 mW.

The SPAD array, consisting of four circular SPADs, is shown in figure 4.17 in more detail. The active area of each HV CMOS SPAD is approximately 5000 μm^2. The SPAD array was illuminated by a single-mode fiber and the height of the fiber end above the chip surface was optimized to maximize the light power incident on the SPADs. The assumption of a Gaussian intensity distribution in the light spot resulted in an effective optical fill factor of 37.7%, i.e. this fraction of the photons exiting the optical fiber fell into the four SPADs in total.

Each SPAD contains a highly doped n+ cathode. The multiplication region is formed below the n+ cathode in the deep p-well. The surrounding deep n-well partially compensates for the deep p-well and, in addition, prevents edge breakdown at the border of the n+ cathode. The p-type epitaxial layer is exploited as the absorption region. The properties of this SPAD structure, such as its DCR, APP,

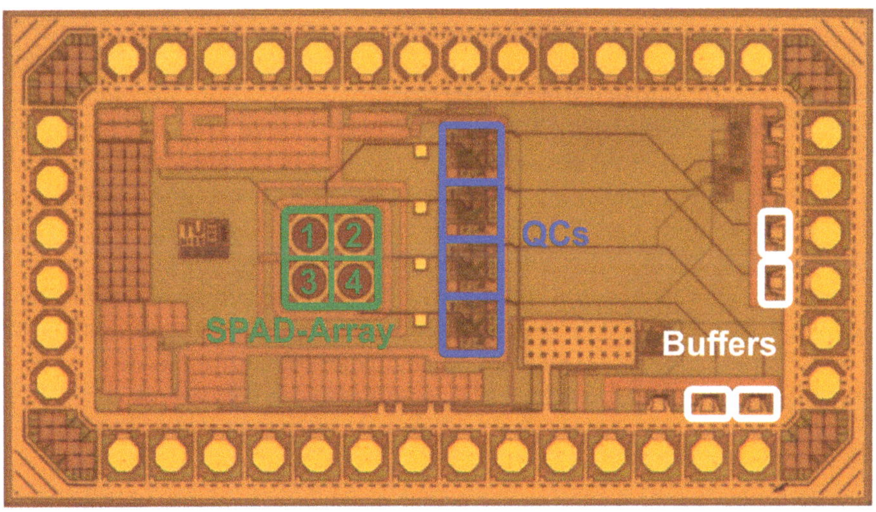

Figure 4.16. Chip photo of the four-SPAD receiver in HV CMOS technology presented in and reproduced from [15] with permission from SPIE.

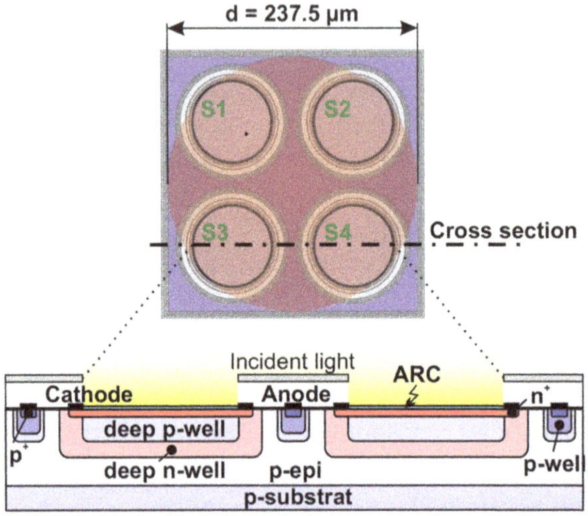

Figure 4.17. Layout of the SPAD array (top) and cross section (bottom) in the HV CMOS four-SPAD receiver. Reproduced from [15] with permission from SPIE.

and PDP, were reported in [17]. The best sample in this article displayed a DCR of 41.7 kcps, an APP of 57.6%, and a PDP of 43.6% at a wavelength of 642 nm (for a dead time of 5.8 ns and an excess bias of 6.6 V).

The parasitic avalanche events, i.e. the dark count rate, afterpulsing probability, and OCTP impose limitations on the achievable sensitivity of SPAD receivers. Therefore, these properties and the breakdown voltage of each SPAD were measured. For these measurements, the dead time was set to 8.9 ns. The breakdown voltages of the four SPADs in this array differed by 0.4 V. SPAD1 (S1) had the lowest breakdown voltage (70.7 V). All SPADs showed similar DCRs (e.g. SPAD3: 82.2 kcps at Vex=6.6 V) and APPs between 35% and 40% for a dead time of 8.9 ns at an excess bias of 6.6 V. In the worst case, i.e. at Vex=6.6 V, the OCTP between two directly neighboring SPADs was, on average, 5.3%. The diagonal crosstalk probability was about 1.8%. Crosstalk between three SPADs was 0.3% at Vex=6.6 V. Optical crosstalk between all four SPADs in the array was not registered within a measurement time of 1 s. The APP of the SPADs strongly depended on the dead time and hold-off time defined by the quenching circuit. In [17], it was shown that by extending the dead time, the APP can be decreased considerably.

A real-time characterization system was used to determine the bit error rate (BER) of the HV four-SPAD receiver. This system included the generation of the laser modulation signal and digital processing by an FPGA for the BER extraction. The output signals of the four quenchers were added digitally at a defined sampling time in the FPGA. The received binary values were then determined using a decision threshold (from one out of four to four out of four) and compared with the original modulation signal by the FPGA. For the best alignment of the optical fiber, the BERs were determined at 50, 100, and 143 Mb s^{-1} by varying the average optical power and the excess bias voltage. The modulation signal for the laser diode was also generated inside the FPGA in order to be able to easily compare the received data with the sent data. Due to the clock rate of the system, which was 1 GHz behind the serializer, it was only possible to generate data signals with a duration of an integer multiple of 1 ns. A bit duration of 7 ns resulted in a data rate of 142.86 Mb s^{-1}, i.e. about 143 Mb s^{-1}. For the modulation signal, a pseudo random bit sequence (PRBS7) and binary RZ coding with a laser duty cycle of 30% were applied at 50 and 100 Mb s^{-1}. At 143 Mb s^{-1}, the duty cycle had to be increased to 43% (3 ns). The BER curves obtained for the HV four-SPAD receiver are displayed in figure 4.18.

In the experiments, the dead time was adapted to the different data rates, to roughly adjust them to the time difference between the falling edge of the RZ signal and the rising edge of the next bit. This led to dead times of 14 ns for 50 Mb s^{-1} and 7 ns for 100 Mb s^{-1}. For 143 Mb s^{-1}, the minimum dead time of 5.8 ns was used. The results presented in figure 4.18 confirm that the critical threshold for forward error correction (FEC) of a BER=2×10^{-3} is clearly underrun for all three data rates. The lowest BERs at these data rates were obtained with a decision threshold of three out of four. At 50 Mb s^{-1}, the sensitivity was −55.1 dBm (3.1 nW). At 100 Mb s^{-1}, the sensitivity was −52.0 dBm (6.4 nW) for a BER of 2×10^{-3}. At 143 Mb s^{-1}, the sensitivity was −38.5 dBm (140.8 nW). When the BER limit was increased to 6.5×10^{-3}, for which forward error correction is still feasible but requires more overhead, the sensitivities were −58.0 dBm (1.6 nW), −54.5 dBm (3.5 nW), and −46.9 dBm (20.3 nW) at 50, 100, and 143 Mb s^{-1}, respectively.

We now discuss these results and compare the PIN-photodiode four-SPAD receivers in the following. To stay below the bit error-correction limits in the presence of the SPAD parasitic avalanche events (in which the APP and OCTP dominate), the receivers described contain arrays of four SPADs. Apart from the Poissonian photon statistics, the sensitivity of SPAD receivers is mainly limited by

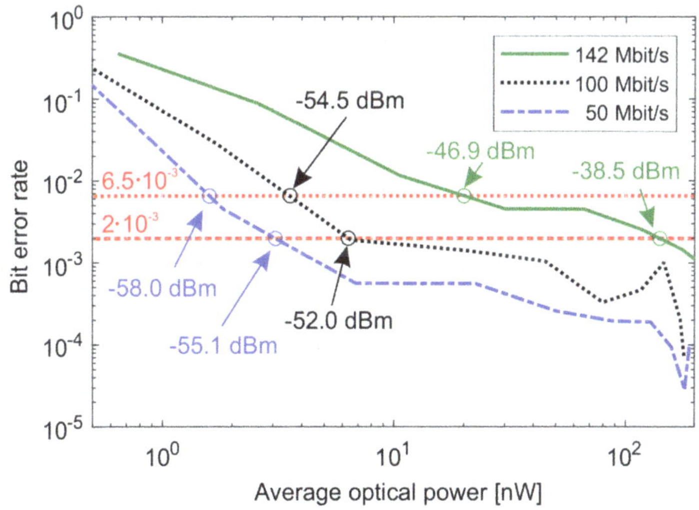

Figure 4.18. Dependence of the BER of the HV CMOS four-SPAD receiver on average optical power. Reproduced from [15] with permission from SPIE.

the optical fill factor, by the photon-number decision threshold, and by the SPADs' PDP. For the geometry used in the HV SPAD receiver, the fill factor was 37.7%. This fill factor results in an optical power loss of 4.2 dB. The decision threshold, which requires that at least three of the four SPADs have to detect a photon, adds a further 4.2 dB to the distance to the quantum limit, if we assume Poissonian photon statistics. This residual gap to the quantum limit mainly arises because the PDP is much smaller than 100%. While these HV CMOS SPADs possess PDPs of up to 43.6% for a 6.6 V excess bias, such a large excess bias voltage results in an overlarge BER, not only because the PDP increases with excess bias, but also because the APP rises with increasing excess bias voltage. The remaining gap to the quantum limit after subtracting the effects of the fill factor and the decision threshold corresponds to the effective PDP of the SPADs at the bias point where the BER limit is reached. The effective PDPs were only 9.5%, 9.3%, and 1.2% at 50, 100, and 143 Mb s^{-1}, respectively, in the experiments with the HV four-SPAD receiver. The BER can be reduced by implementing a larger number of SPADs in the array and by increasing the decision threshold (i.e. the number of SPADs that have to detect a photon during the same bit in order to receive a digital '1') or by increasing the dead time. With a higher decision threshold, the influence of the APP decreases. However, a higher decision threshold also increases the resulting loss in optical power. An increased dead time directly decreases the APP but limits the achievable data rate.

For a four-SPAD receiver fabricated in PIN-photodiode CMOS [5], the best sensitivities were achieved with analog processing of the quencher output signals in MATLAB (-55.7 and -51.6 dBm at 50 and 100 Mb s^{-1}, respectively, at BER=2×10^{-3} and with a dead time of 9 ns). This analog processing approach applied a gliding filter, which was optimized for each data rate. When a digital processing approach was used for the quencher output signals, the sensitivity in RZ (BER=2×10^{-3}) at 100 Mb s^{-1} was −47.8 dBm in [5] and −46.3 dBm in [14]. Compared to the digital processing approaches with the SPAD receivers in PIN-photodiode CMOS technology presented in [5] and [14], the HV receiver described here achieved an improvement of the sensitivity to −52.0 dBm at 100 Mb s^{-1} (BER=2×10^{-3}). A BER of 2×10^{-3} was impossible at 150 Mb s^{-1} in [14], whereas the HV CMOS four-SPAD receiver reached this bit error rate at 143 Mb s^{-1}. A large advantage of the SPADs in the HV CMOS four-SPAD receiver is that they support an opto-window with an ARC, which the SPADs in the PIN-photodiode receiver do not have. Without the opto-window, interference effects within the oxide and passivation stack can, depending on the wavelength, cause increased reflection and therefore a larger loss of incident light. The spectral PDP then contains 'oscillations.' This is not the case for SPADs with an opto-window. The advantage of the PIN-photodiode CMOS process is its thick lightly doped epitaxial layer, which has much less doping than that used in HV CMOS technology. Because of this lower epitaxial doping, a thick depletion zone is already available at smaller reverse voltages than those of the HV CMOS process. Therefore, if a thick depletion zone is required, the operating voltage of SPADs in the HV CMOS process needs to be higher, which is particularly needed for wavelengths in the near-infrared range. However, the HV transistors that are only available in HV CMOS potentially allow the integration of quenchers with even higher excess bias voltages, leading to a higher PDP.

The results reported in [15] show that the high red-light PDP of SPADs in a standard HV CMOS technology without process modifications enables a receiver to have better sensitivities than two four-SPAD receivers in PIN-photodiode CMOS using comparable digital quencher output processing. The HV CMOS four-SPAD receiver in a cheap sub-micrometer technology achieved 143 Mb s^{-1} with a BER of even less than 2×10^{-3}, enabling effective forward error correction. For the HV CMOS four-SPAD receiver, the distances to the quantum limit for BER = 2×10^{-3} at 50 and 100 Mb s^{-1} are 18.6 dB and 18.7 dB, respectively (the quantum limit is at −73.7 and −70.7 dBm for 50 and 100 Mb s^{-1}, respectively).

A gating SPAD receiver was introduced in [18]. This receiver used five sub-bits within a bit to be able to receive the data with only one SPAD. The sub-bit principle is shown in figure 4.19 for four sub-bits. If a photon is detected in the gate phase (V_{EX}, i.e. using a high voltage at the SPAD to make it sensitive), the counter counts up by one. In the designed circuit (see figure 4.20), five sub-bits were realised as a shift register and a digital comparator (circuit block Thrh) with selectable thresholds of two out of five, three out of five, four out of five, and five out of five for a decision on a logical ‘1’ [18].

The gating receiver uses cascoded switches (N0 and cascode transistor N1 connect the SPAD to V_{SS} = −3.3 V, P0 and cascode transistor P1 connect the SPAD to V_{SPAD} = +3.3 V). The excess bias voltage used for the SPAD therefore can be up to 6.6 V. During the active phase, when the SPAD is connected to V_{SPAD}, the node PLS charges toward V_{SPAD}; when V_{SPAD} is almost reached, P0 is switched off by the block SPAD control. When a photon (or a dark count or an afterpulse) triggers an avalanche, the SPAD discharges node PLS to about 0 V, limited by P1; the node CAT is discharged to V_{SS}, i.e. if V_{SUB} is properly set to the breakdown voltage V_{BR} across the SPAD, it quenches the SPAD. The comparator can detect an avalanche via the transmission gate (the state of PLS is dynamically stored at PLSSH during the next reset phase). The pulse sequence is written into a five-bit shift register controlled by CKLS via a buffer consisting of inverters.

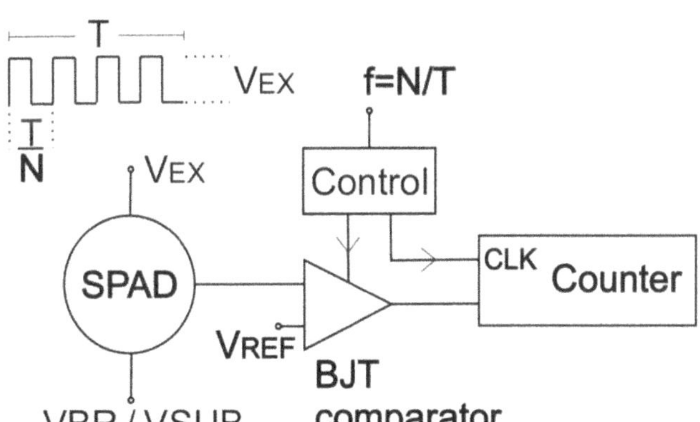

Figure 4.19. Principle of a gating receiver that uses sub-bits.

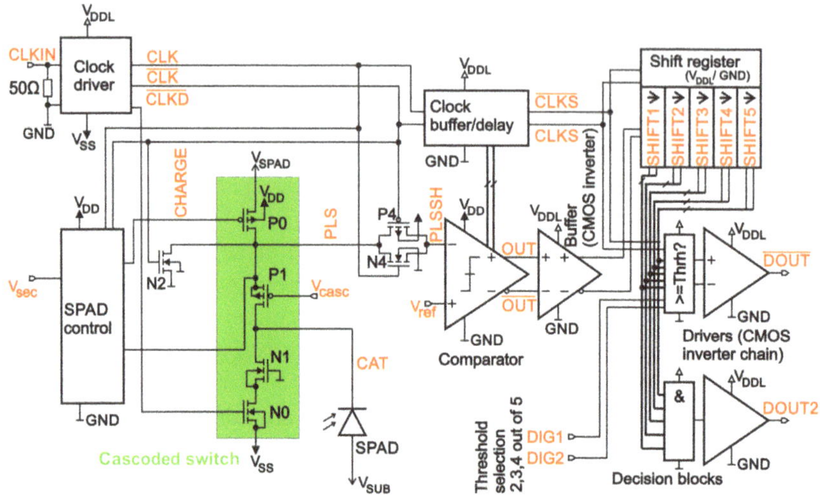

Figure 4.20. Block diagram of a gating receiver that uses sub-bits © 2018 IEEE. Reprinted, with permission, from [18].

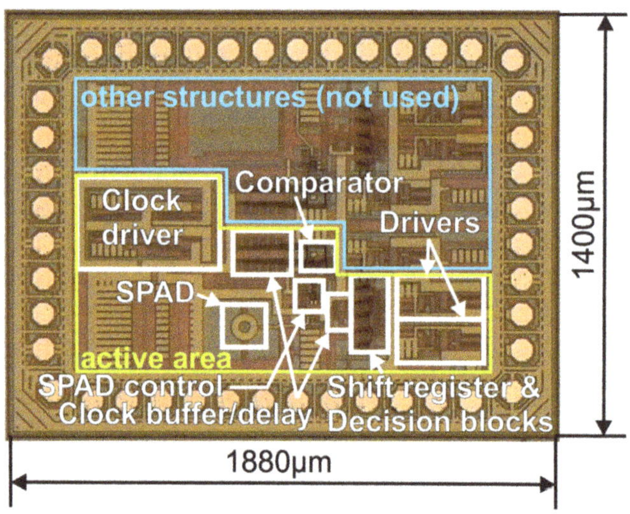

Figure 4.21. Chip photo of the gating SPAD receiver that uses sub-bits. © 2018 IEEE. Reprinted, with permission, from [18].

If no avalanche occurred during the active phase (the SPAD is charged close to V_{SPAD}), the SPAD is quenched during the reset phase, when N0 is switched on.

The digital inputs DIG1 and DIG2 define which threshold (two out of five, three out of five, four out of five, or five out of five) is used to obtain an output signal at DOUT. DOUT2 only switches to a logical '1' when five events are counted within one bit.

A microphotograph of the fabricated gating receiver is shown in figure 4.21. In fact, the gating receiver has an active area of 0.66 mm^2 and occupies only about half of the total active chip area. The total dimensions of the chip are $1880 \times 1400\,\mu m^2$ [18]. The cascoded gating receiver is integrated together with a thick SPAD in $0.35\,\mu m$ PIN-photodiode CMOS that has a diameter of about $50\,\mu m$.

The results of bit error measurements at $20\,Mb\,s^{-1}$ in NRZ with PRBS7 are shown in figure 4.22. At $20\,Mb\,s^{-1}$, the gating receiver achieved a sensitivity of −64 dBm using 635 nm light with on-off keying (NRZ), for which a threshold of two photons for a logical '1' was sufficient. The extinction ratio of the light source was high (larger than 100). The results of bit error measurements at $50\,Mb\,s^{-1}$ in NRZ with PRBS7 are shown in figure 4.23. At $50\,Mb\,s^{-1}$, the sensitivity was −57 dBm, also using a threshold of two photons for a logical '1' [18].

The clock frequency has to be five times the data rate, which limits the maximum data rate of this sub-bit one-SPAD receiver. However, there is no fill-factor influence, unlike the case of the four-SPAD receiver described above. In addition, optical crosstalk between the SPADs does not worsen the BER in the sub-bit one-SPAD receiver. This enables the gating receiver to have better sensitivity.

A 64 × 64 SPAD receiver was introduced using $0.13\,\mu m$ CMOS image-sensor technology [10]. The large number of SPADs allowed dead times to be applied for longer than a bit period to reduce the afterpulsing probability and thereby the bit error rate at higher data rates. Thin SPADs with passive quenching and digital signal processing were implemented. An on-off-keying data rate of $400 Mb\,s^{-1}$ with a sensitivity of −49.9 dBm using 450 nm light was reported. A data rate of $500\,Mb\,s^{-1}$ was reported for 4-level pulse amplitude modulation (4-PAM) transmission with a minimum optical power of −46.1 dBm. The electrical power consumption was 230 pJ/bit.

Figure 4.24 compares the sensitivities of SPAD receivers. The 64 × 64 SPAD receiver reported in [10] comes closest to the quantum limit, with distances of

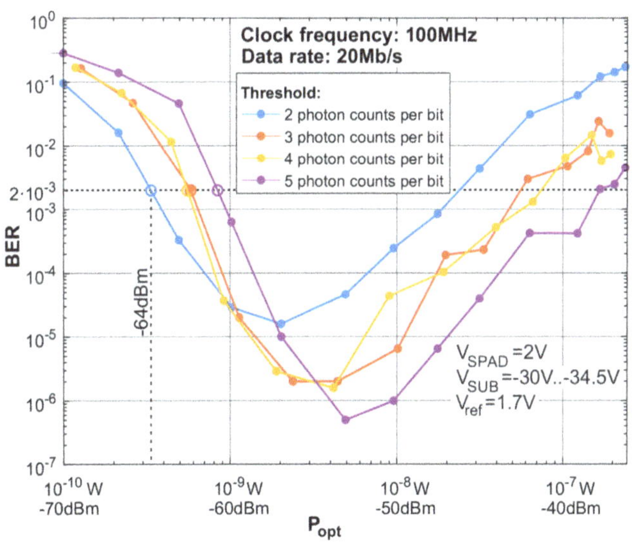

Figure 4.22. Dependence of the BERs of the gating SPAD receiver using sub-bits on average optical input power at $20\,Mb\,s^{-1}$. © 2018 IEEE. Reprinted, with permission, from [18].

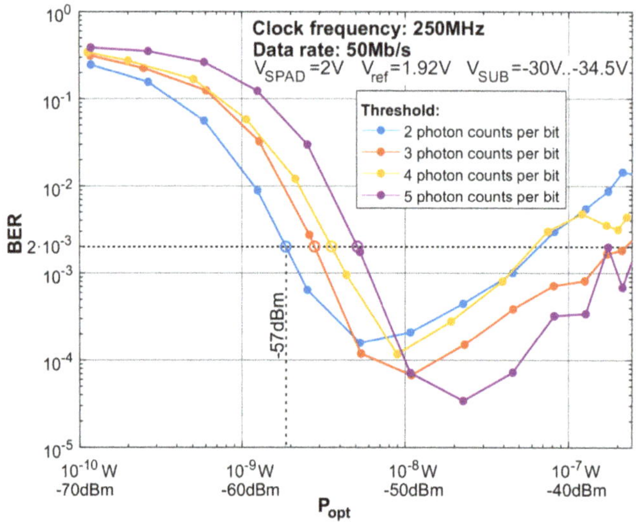

Figure 4.23. Dependence of the BERs of the gating SPAD receiver using sub-bits on average optical input power at 50 Mb s^{-1}. © 2018 IEEE. Reprinted, with permission, from [18].

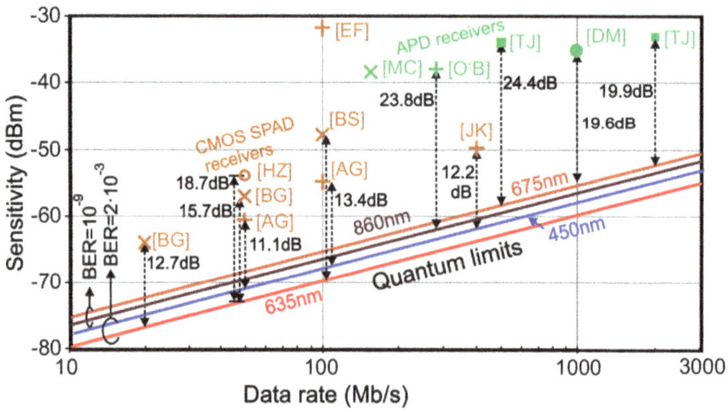

Figure 4.24. Comparison of SPAD receivers' sensitivities (brown data: [AG]=[19], [BG]=[18], [BS]=[14], [HZ]=[5], [JK]=[10]) and APD receivers' sensitivities (green data: [DM=[2]], [MC]=[20], [O'B]=[21], [TJ]=[22, 23]) with distances to the quantum limit.

11.1 dB at 50Mb s^{-1} and 13.4 dB at 100 Mb s^{-1} for 450 nm wavelength light [19]. SiPM receivers have also been reported [24, 25] with sensitivities of −53.4 dBm at 400Mb s^{-1} (8.7 dB above the quantum limit) and −49 dBm at 1 Gb s^{-1} for 405 nm light.

4.3 Optical wireless communications experiments with SPAD receivers

The better sensitivity of SPAD receivers compared to APD linear-mode receivers suggests that they could also be used in OWC to extend the transmission distance or to reduce the transmitter power. Unless otherwise noted, the OWC experiments described in the following were performed in a normally lighted laboratory with an illuminance of about 500 lx. We used the four-SPAD receiver described in [5] (see figure 4.9) as the receiver in an OWC experiment together with a 635nm CW laser and an external modulator (to achieve a high extinction ratio). This light source was coupled to a single-mode fiber and a Thorlabs collimator F280FC-B to form a light beam with a 0.01° divergence angle [26]. The four-SPAD receiver was placed in a black box with a small hole covered by an optical interference filter with a bandwidth of 10 nm. The distance between transmitter and receiver was 2 m. The dependence of the BER results on the optical output power of the modulated laser at 50 Mb s^{-1} in NRZ is shown in figure 4.25. A very small optical output power of less than 1 μW was sufficient to achieve a BER below the error-correction limit.

For the next OWC experiments, we used a 650 nm resonant-cavity (RC) LED with an aspherical lens in front of it. A mirror was used to increase the transmission distance within the dimensions of the laboratory. The setup can be seen in figure 4.26 [27]. The output power of the RC LED was 1.1 mW and a collimator produced a beam divergence of 0.038 rad.

The BERs obtained for distances of up to 6 m are shown in figure 4.27. The maximum OWC distance for a BER below the BER limit at 50 Mb s^{-1} in NRZ is 5.3 m. The dependence of the BER on the background light level can be seen in figure 4.28. The BER is less than 1.9×10^{-3} for ambient light illuminances of up to 2 klx [27].

Since RZ was found to produce a lower BER than NRZ in [5], RZ was also used and the duty ratio was varied at a distance of 3m [28]. The data rate could be increased to 75 Mb s^{-1}. Figure 4.29 depicts the BER results achieved. Duty ratios of

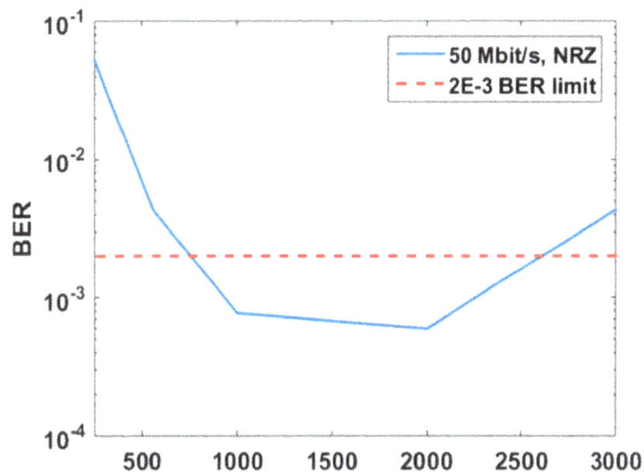

Figure 4.25. Dependence of the BER of OWC over 2 m with a four-SPAD receiver at 50 Mb s^{-1} on average optical transmitter power. © 2017 IEEE. Reprinted, with permission, from [26].

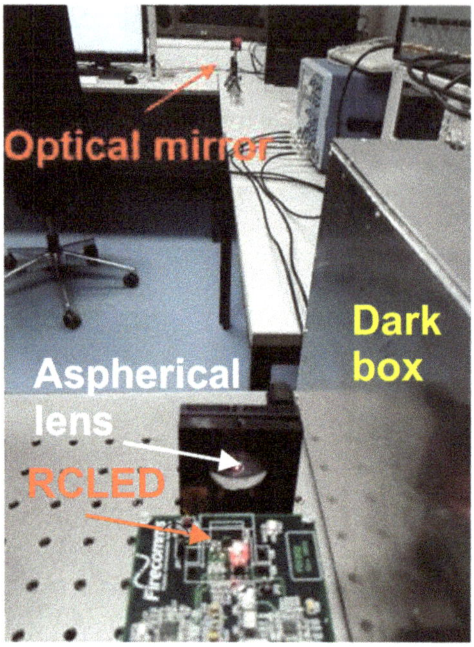

Figure 4.26. Setup for OWC experiments with a mirror to double the transmission distance. © 2018 IEEE. Reprinted, with permission, from [27].

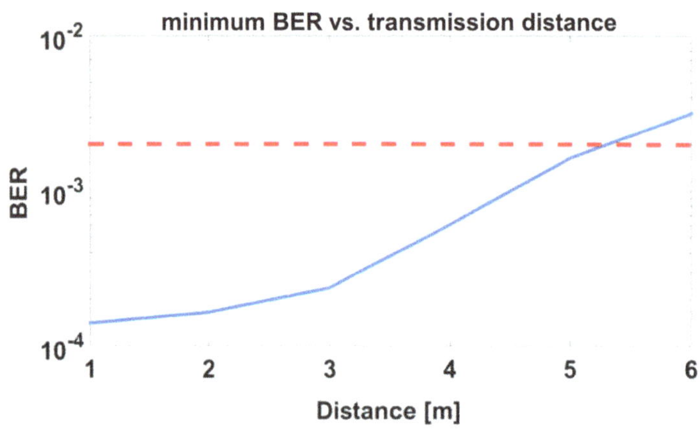

Figure 4.27. Dependence of the bit error ratio of the OWC experiments with the RC LED on distance. © 2018 IEEE. Reprinted, with permission, from [27].

around 50% led to the lowest BERs. The much larger increase of the BER at higher duty ratios for 75 Mb s^{-1} may be caused by a slow tail in the transient of the light emitted by the RC LED.

A single pixel of a GaN microLED that emitted at 450 nm was used together with the 64 × 64 SPAD receiver [10] in a free-space experiment over 50cm [19]. In addition, a neutral-density filter and a collection lens were used to focus the light

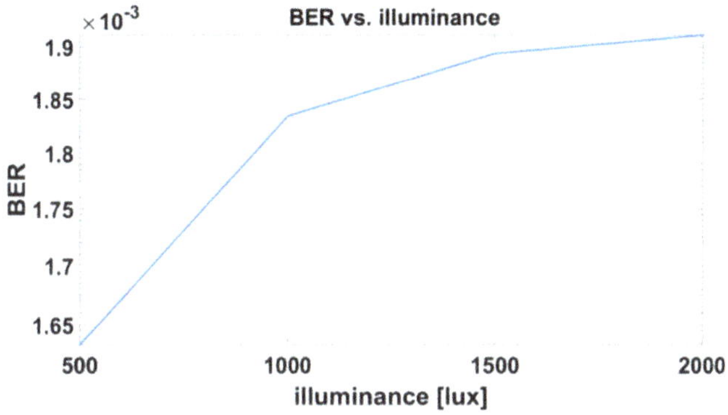

Figure 4.28. Dependence of the bit error ratio of the OWC experiments with an RC LED at a 5 m distance on background illumination. © 2018 IEEE. Reprinted, with permission, from [27].

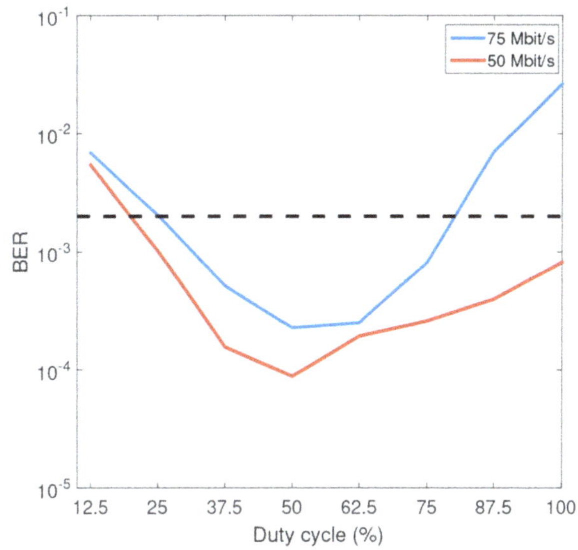

Figure 4.29. Dependence of the bit error ratio of the OWC experiments with an RC LED at a 3 m distance on the duty ratio. © 2018 IEEE. Reprinted, with permission, from [28].

onto the light-sensitive area of the SPAD receiver. Receiver sensitivities of −60.5 dBm at 50 Mb s^{-1} and −55.2 dBm at 100 Mb s^{-1} were reported for 450 nm light, corresponding to 41 and 68 incident photons per bit, respectively [19].

References

[1] Ebeling K J 1993 *Integrated Optoelectronics* (Berlin: Springer)

[2] Milovancev D, Brandl P, Jukic T, Steindl B, Vokic N and Zimmermann H 2019 Optical wireless APD receivers in 0.35 μm HV CMOS technology with large detection area *Optics Exp.* **27** 11930–45

[3] Haigh F A 1967 *Handbook of the Poisson Distribution* (New York: Wiley)
[4] Mahmoudi H, Hofbauer M, Steindl B, Schneider-Hornstein K and Zimmermann H 2018 Modeling and analysis of BER performance in a SPAD-based integrated fiber optical receiver *IEEE Photon. J.* **10** 7908411
[5] Zimmermann H, Steindl B, Hofbauer M and Enne R 2017 Integrated fiber optical receiver reducing the gap to the quantum limit *Sci. Rep.* **7** 2652
[6] Mahmoudi H, Hofbauer M, Steindl B, Schneider-Hornstein K and Zimmermann H 2019 Statistical study of intrinsic parasitics in an SPAD-based integrated fiber optical receiver *IEEE Trans. Electron Dev.* **66** 497–504
[7] Ramilli M, Allevi A, Chmill V, Bondani M, Caccia M and Andreoni A 2010 Photon-number statistics with silicon photomultipliers *J. Opt. Soc. Amer. B, Opt. Phys.* **27** 852–62
[8] Gallego L, Rosado J, Blanco F and Arqueros F 2013 Modeling crosstalk in silicon photomultipliers *J. Instrum.* **8** P05010
[9] Savuskan V, Brouk I, Javitt M and Nemirovsky Y 2013 An estimation of single photon avalanche diode (SPAD) photon detection efficiency (PDE) nonuniformity *IEEE Sensors J.* **13** 1637–40
[10] Kosman J, Almer O, Abbas T A, Dutton N, Walker R, Videv S, Moore K, Haas H and Henderson R 2019 29.7 A 500 Mb/s-46.1 dBm CMOS SPAD receiver for laser diode visible-light communications *IEEE Int. Solid-State Circuits Conf.* pp 468–70
[11] Fisher E, Underwood I and Henderson R 2013 A reconfigurable single-photon counting integrating receiver for optical communications *IEEE J. Solid-State Circ.* **48** 1638–50
[12] Sklar B 2001 *Digital Communications: Fundamentals and Applications* (Englewood Cliffs, NJ: Prentice-Hall)
[13] ITU-T 2004 G.975.1: Forward error correction for high bit-rate dwdm submarine systems *Telecommunication Standardization Sector* E 27093 ITU
[14] Steindl B, Hofbauer M, Schneider-Hornstein K, Brandl P and Zimmermann H 2018 Single-photon avalanche photodiode based fiber optical receiver up to 200 Mb/s *J. Sel. Topics Quantum Electron.* **24** 3801308
[15] Hofbauer M, Steindl B and Zimmermann H 2020 Fully integrated optical receiver using single-photon avalanche diodes in high-voltage CMOS *Optical Eng.* **59** 070502
[16] Hofbauer M, Steindl B, Schneider-Hornstein K and Zimmermann H 2019 Thick CMOS single-photon avalanche diode optimized for near infrared with integrated active quenching circuit *Single-Photon Workshop (SPW)* Milano
[17] Hofbauer M, Steindl B, Schneider-Hornstein K and Zimmermann H 2020 Performance of high-voltage CMOS single-photon avalanche diodes with and without well-modulation technique *Optical Eng.* **59** 040502–8
[18] Goll B, Hofbauer M, Steindl B and Zimmermann H 2018 A fully integrated SPAD-based CMOS data-receiver with a sensitivity of -64 dBm at 20 Mb/s *IEEE Solid-State Circ. Lett.* **1** 2–5
[19] Griffith A D, Herrnsdorf J, Almer O, Henderson R K, Strain M J and Dawson M D 2019 High-sensitivity free space optical communications using low size, weight and power hardware 1902.00495v1 1–7
[20] McCullagh M J and Wisely D R 1994 155 Mbit/s optical wireless link using a bootstrapped silicon APD receiver *IET Electron. Lett.* **30** 430–2
[21] O'Brien D *et al* 2012 High-speed optical wireless demonstrators: conclusions and future directions *J. Lightwave Technol.* **30** 2181–7

[22] Jukić T, Steindl B, Enne R and Zimmermann H 2016 200 μm APD OEIC in 0.35 μm BiCMOS *IET Electron. Lett.* **52** 128–30

[23] Jukić T, Steindl B, Enne R and Zimmermann H 2016 400 μm diameter APD OEIC in 0.35 μm BiCMOS *IEEE Photonics Technol. Lett.* **28** 2004–7

[24] Ahmed Z, Zhang L, Faukner G, O'Brien D and Collins S 2019 A shot-noise limited 420 Mbps visible light communication system using commercial off-the-shelf silicon photomultiplier (SiPM) *IEEE Int. Conf. Commun. Workshops (ICC Workshops)* pp 1–5

[25] Ahmed Z, Singh R, Ali W, Faukner G, O'Brien D and Collins S 2020 A SiPM-based VLC receiver for Gigabit communication using OOK modulation *IEEE Photonics Technol. Lett.* **32** 317–20

[26] Milovančev D, Jukić T, Steindl B, Hofbauer M, Enne R, Schneider-Hornstein K and Zimmermann H 2017 Optical wireless communication with monolithic avalanche photodiode receivers *2017 IEEE Photonics Conference (IPC)* pp 25–6

[27] Milovančev D, Weidenauer J, Steindl B, Hofbauer M, Enne R and Zimmermann H 2018 Visible light communication at 50 Mbit/s using a red LED and an SPAD receiver *2018 11th Int. Symp. on Communication Systems, Networks Digital Signal Processing (CSNDSP)* pp 1–4

[28] Milovančev D, Weidenauer J, Steindl B, Hofbauer M, Enne R and Zimmermann H 2018 Influence of on-off keying duty cycle on BER in wireless optical communication up to 75 Mbit/s using an SPAD and a RC LED *Int. Conf. on Broadband Communications for Next Generation Networks and Multimedia Applications (COBCOM)* pp 1–5

Michael Hofbauer, Kerstin Schneider-Hornstein and Horst Zimmermann

Chapter 5

SPADs in quantum applications

Due to fast progress in the improvement of the performance of photon sources, detectors, and structures for manipulating the states of photons, photonic systems have emerged in countless quantum applications, such as quantum key distribution, quantum computing, quantum simulation, ghost imaging, and super-resolution microscopy. In this chapter, we will explain the basic concepts of these applications and discuss selected examples, as well as the most critical requirements for photodetectors in these applications. Additionally, the key properties of single-photon avalanche diodes (SPADs) will be compared to those of superconducting nanowire single-photon detectors (SNSPDs), which are highly sensitive photodetectors that are operated at cryogenic temperatures.

5.1 Introduction

In 2020, the first photonic quantum simulator reached quantum supremacy, meaning that the outcome of an experiment is not simulatable using classical computing in a reasonable amount of time [1]. For today's photonic quantum computing, the most difficult task is the construction of two-input gates, due to the bosonic character of photons. However, solutions even exist for this problem. These solutions require highly effective single-photon detection [2], which is currently only guaranteed by superconducting single-photon detectors (SNSPDs).

Among the most prominent photonic quantum applications, quantum key distribution (QKD) has already started to leave its experimental status and find its way into the communications market, which has also been fuelled by the immanent threat that quantum computing might render many modern encryption techniques useless in the near future. The main challenge to the currently available fiber-based QKD systems is a range limitation of approximately 200 km for reasonable key rates [3]. The current record distance is 509 km; however, the secure key rate per signal pulse is only 6.19×10^{-9} [4]. In order to continue the use of

doi:10.1088/978-0-7503-2584-4ch5

secure communications over extended distances without the use of trusted nodes, satellite systems, most of them in low Earth orbit (LEO), are currently being evaluated [5].

Other promising applications are ghost imaging and super-resolution microscopy. Both exist in flavours that do not need any quantum effects, as well as ones that benefit from such effects. In ghost imaging, a pixel detector and a large 'bucket' detector are typically used, and the image can be reconstructed even though the light that reaches the pixel detector never has interacted with the object to be imaged [6].

Super-resolution microscopy, i.e. microscopy that allows imaging below the Abbe limit, uses two different approaches that will be discussed in this chapter. The first is a classical approach, in which switchable point light sources can be attached to the structure to be measured. The positions of these point light sources are then estimated by Gaussian fits of the detected photon clouds. A large range of related techniques has evolved and resolutions down to a few tens of nanometers have been demonstrated [7]. The second super-resolution microscopy approach utilizes the fact that if entangled photons are used to illuminate a sample, the achievable resolution can theoretically be reduced by a factor of N if a photon source with N-fold entangled photons is used [8].

For most of the abovementioned applications, a high photon detection efficiency is crucial. However, each of the following sections will discuss which single-photon avalanche diode (SPAD) parameters are the most critical for a particular application and how imperfections influence the systems' performance and scalability.

5.2 Superconducting nanowire single-photon detectors

Superconducting nanowire single-photon detectors (SNSPDs) are very promising detectors, which are capable of detecting single photons very efficiently over a broad wavelength range from ultraviolet (UV) up into the infrared (IR) wavelength range, including the wavelength range around 1550 nm, which is very important for telecoms [9].

In this section, we will first discuss the basic operating principle of SNSPDs. In the subsequent part, the key parameters of a single-photon detector will be discussed, and we then compare these key parameters for SNSPDs and SPADs.

An SNSPD consists of a thin (typically 100–200 nm wide [9]) superconducting wire that is biased only slightly below its critical current, as indicated in figure 5.1. Above this critical current, the material leaves the superconducting state and becomes resistive again.

The basic principle of the SNSPD is explained in e.g. [9], and briefly summarized in this paragraph. Figure 5.2 depicts the different states of the nanowire during the detection process. If a photon is absorbed, the temperature in the absorption region rises above the critical temperature. Consequently, the material becomes locally resistive. The current must then flow around this resistive region, which increases the current density in the region. Since the SNSPD is operated only slightly below its critical current density, this increased current density results in a growing resistive region, until the border of the nanowire is reached. The power dissipated in this

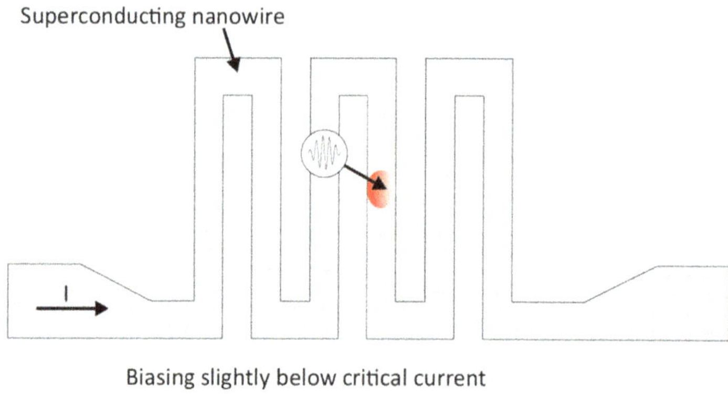

Figure 5.1. Superconducting nanowire: basic principle.

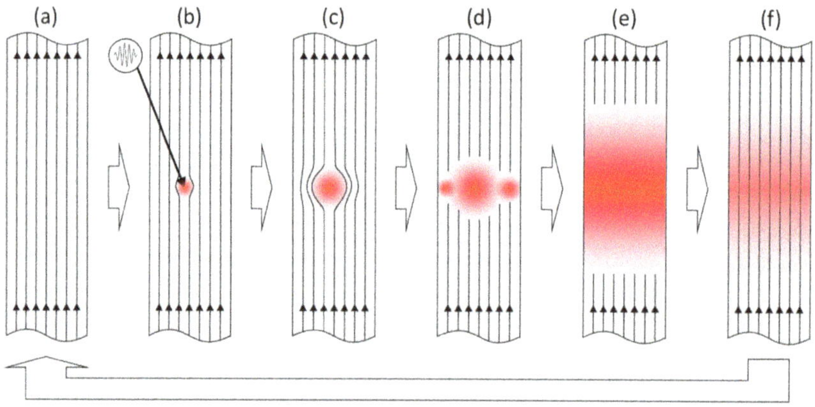

Figure 5.2. Detection process in a superconducting nanowire after [9–12]: (a) biased slightly below the critical current. (b) Photon absorption causes a resistive hot spot that forces the superconducted current to flow around it. (c) Due to the increased current density at the narrow points, a wider fraction of the wire becomes resistive. (d) At some point, the whole width is affected and the nanowire becomes resistive. (e) The resistive region grows along the axis of the nanowire due to Joule losses until the bias current is deviated through a shunt resistor. (f) The nanowire cools down and becomes superconductive again.

resistive part of the nanowire then leads to further growth of the resistive region along the axis of the nanowire. The voltage transient across the then partly resistive nanowire can be measured; this indicates that a photon was detected. As soon as this voltage drop is detected, the nanowire can be reset by drawing the current through a bypass, which allows the nanowire to cool down and become superconductive again.

5.2.1 Key parameters of a single-photon detector

To compare SPADs with superconducting nanowire single-photon detectors (SNSPDs), we will first take a closer look at the key parameters that define the quality of a single-photon detector from a user's point of view. We will not discuss the reasons for imperfections and parasitic effects in detail, but concentrate more on the implications for the sensor's output. We expect the behaviour depicted in figure 5.3 from an ideal single-photon detector. As soon as a photon is detected, a readable (i.e. sufficiently large) output signal is generated without any delay. For the ideal detector, each photon generates an output pulse, and no pulse is generated if no photon is present.

However, real single-photon detectors suffer from parasitic effects as well as from imperfections. One of these imperfections is a limited photon detection efficiency (PDE). The PDE is defined as [13]:

$$PDE = \frac{N_{\text{detphot}}}{N_{\text{phot}}}, \tag{5.1}$$

where N_{detphot} is the number of detected photons and N_{phot} is the total number of photons incident on the detector. Figure 5.4 depicts the output signal of a detector with a PDE smaller than 1. A detailed model for the PDE of SPADs is presented in section 1.4. The PDE also includes losses due to a limited fill factor or reflections at the surface of the detector. The term 'photon detection probability' (PDP) is used for

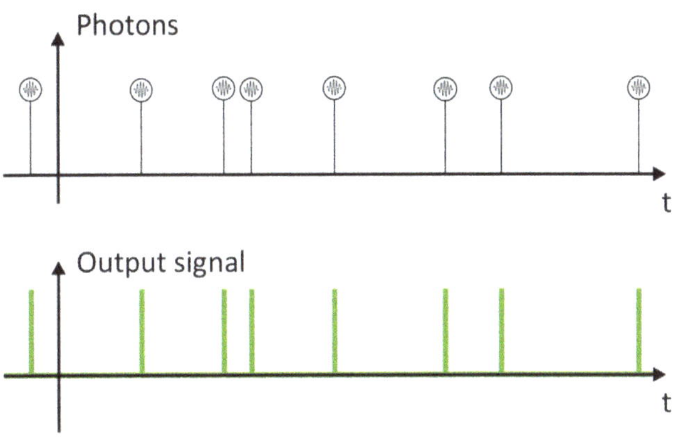

Figure 5.3. Output signal of an ideal single-photon detector.

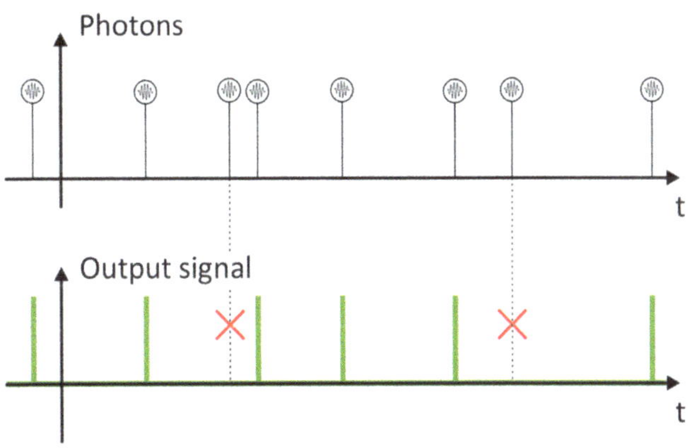

Figure 5.4. Output signal of a single-photon detector with non-ideal PDE.

the number of pulses produced compared to the number of photons entering the active area of the photodetector. An ideal single-photon detector would have a PDP equal to 1 (i.e. a probability of 100%).

Another imperfection of photodetectors is their dark count rate (DCR). Even if no photons are incident at the single-photon detector, it can still generate an output pulse, as depicted in figure 5.5. For SPADs, this can be caused either by thermal generation or by tunnelling effects [14]. The DCR is defined as follows:

$$DCR = \frac{N_{\text{counts}}}{t_{\text{meas}}}, \tag{5.2}$$

where N_{counts} is the total count when no photons are incident at the detector and t_{meas} is the total measurement time. An ideal single-photon detector has a DCR of 0 counts per second (cps).

The afterpulse is another parasitic effect that is very common in single-photon detectors. An afterpulse is an output pulse that is not triggered by a photon; rather, it is a side effect of a previous detection. In SPADs, such afterpulses are caused, e.g. by traps that are filled during an avalanche and released shortly afterwards [14]. The charge released from the trap can then trigger another avalanche. An example of such behaviour is depicted in figure 5.6.

The afterpulsing probability (APP) is defined as follows:

$$APP = \frac{N_{\text{AP}}}{N_{\text{counts}}}, \tag{5.3}$$

where N_{AP} corresponds to the number of afterpulses. For an ideal single-photon detector, the APP is 0%.

The dead time after each detection constitutes an additional imperfection. As depicted in figure 5.7, the output signal cannot show another detection while the pulse of the previous detection is still active [13]. In general, the dead time and the pulse width of the output signal are not necessarily connected. However, in most

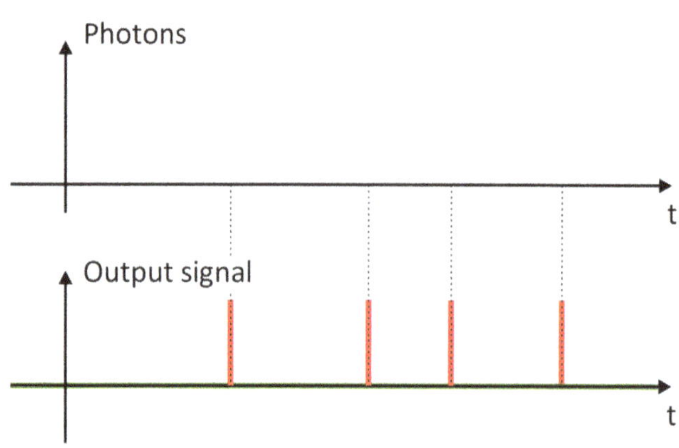

Figure 5.5. Output signal of a single-photon detector showing dark counts.

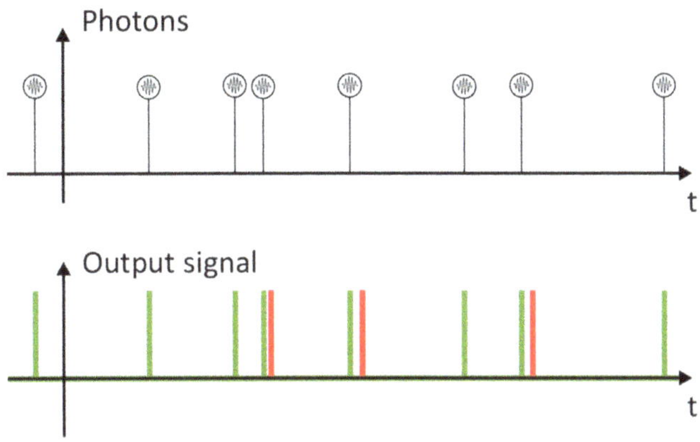

Figure 5.6. Output signal of a single-photon detector showing afterpulsing.

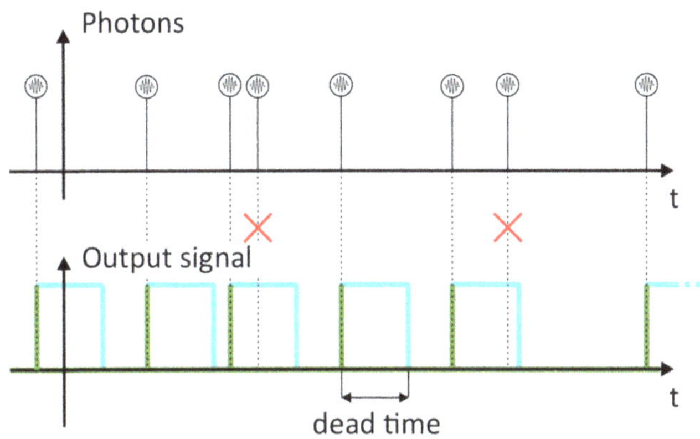

Figure 5.7. Output signal of a single-photon detector with a non-zero dead time.

cases, the dead time will be as long as, or longer than, the output pulse. An ideal single-photon detector has a dead time of 0 s.

Finally, for an ideal detector, we expect the output pulse to occur simultaneously with the photon detection. A constant delay between both is acceptable for most applications. However, the delay between the incoming photon and the output pulse is typically not constant, but shows a statistical distribution, as depicted in figure 5.8. This timing jitter can be statistically characterized. Very often, the full width at half maximum (FWHM) value of the statistical distribution of the delay time is used as characteristic value in comparisons of different single-photon detectors [13]. Ideally, the timing jitter is 0 s.

5.2.2 A comparison of SPADs and SNSPDs

In this part, we compare the performances of SPADs and SNSPDs. The main advantage of SNSPDs is their superior PDE, especially in the telecoms wavelength range. Their main drawback is their cryogenic operating temperatures, typically around 2 K or even lower.

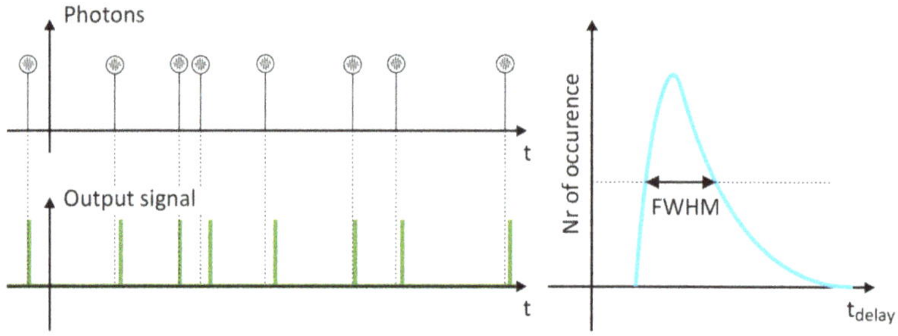

Figure 5.8. Output signal of a single-photon detector showing timing jitter (left) and a delay-time histogram (right).

5.2.2.1 Photon detection efficiency

While commercial counting modules with silicon SPADs reach decent peak PDEs of typically up to approximately 70% [15, 16] (in [17], one model even reaches typical values of 75%), they typically achieve their peak PDEs for red light (i.e. at around 630 nm). Recent integrated SPADs using complementary metal–oxide–semiconductor (CMOS) technology have also achieved peak PDEs very close to 70%, as presented in [18, 19]. Integration in CMOS has the big advantages that the quenching circuit can be implemented on the same chip and that it is possible to integrate additional intelligence on the same chip, such as coincidence measurements, time-stamping circuitry, and many more.

However, for many applications such as QKD, single-photon detectors with high sensitivities at around 1550 nm are required in order to use available telecoms infrastructure, such as optical fiber links. Additionally, at these wavelengths, optical fibers such as SMF28 are available, which have a considerably lower loss of $\sim$0.2 dB km^{-1} [20] compared to $\leqslant$ 10 dB km^{-1} at 630 nm for an S630-HP fiber [21]. In this wavelength range, silicon is transparent due to its bandgap of $\sim$1.12eV [22],[1] and therefore, silicon SPADs cannot be directly used to detect single photons at 1550 nm.

Commercially available SNSPD systems, such as IDQ's ID281 reach system detection efficiencies of up to 90% [23] over a broad wavelength range, which also includes 1550 nm. In 2020, Reddy *et al* published details of an SNSPD with a system detection efficiency (including the coupling of light from a fiber to the detector) of 98% [24]. They utilized a vertical distributed Bragg reflector (DBR) as a mirror below a meandered nanowire. In addition, they used relatively large active areas of up to 50 μm in diameter. The reported dark count rate was in the range of a few thousand counts/s and the timing jitter ranged from $\sim$500 ps to 4 ns, depending on the series resistor used for the 50 μm device.

Compared to this high PDE, the best SPADs in this wavelength range (typically InGaAs/InAlAs SPADs) reach PDEs in the range of 20%–30%. In 2021, Zhang *et al* presented an InGaAs/InAlAs SPAD that reached a PDE of 36% [25]. Its DCR was approximately 1.9×10^7 cps, which was considerably higher than those of the best

[1] This bandgap energy corresponds to a cutoff wavelength of $\sim$1.11 μm.

SNSPDs. However, these SPADs were operated at 240 K, while SNSPDs are operated at cryogenic temperatures, typically at around 2 K or even lower.

5.2.2.2 Detector noise

In terms of detector noise, SNSPDs normally outperform SPADs under typical operating conditions. However, this comparison is not completely fair, since a considerable part of the DCR in SPADs originates from thermal generation [14], and SPADs are typically operated at much higher temperatures than SNSPDs.

The SNSPD described in [26], which was optimized for detection in the UV range, had a dark count rate of only 0.25 counts per hour. In SNSPDs optimized for longer wavelengths, e.g. the very important wavelength range around 1550 nm, the DCR is typically higher. In [27], an SNSPD with a DCR of 0.5 cps was presented, which still achieved a PDE of 80%. The authors achieved this low DCR by integrating a 40 nm optical bandpass filter at the fiber end that coupled light into the SNSPD, filtering out part of the black-body radiation of the fiber.

Commercial counting modules that use silicon SPADs also achieve low DCRs; e.g. a module from Laser Components reaches a DCR as low as 10 cps [15].

InGaAs SPADs for 1550 nm typically have much higher DCRs. For example, the SPAD described in [25] suffered from a DCR of 1.9×10^7 cps and had a high PDE. However, the commercial counting module ID230 from IDQ has a DCR of ~50 cps [28]. The authors of [29] even achieved a DCR as low as around 7 cps at a reduced excess bias voltage and a reduced temperature of $-110\ ^\circ C$. At this low temperature, the DCR even stayed below 30 cps at a higher excess bias of 3.5 V.

Afterpulsing seems not to be pronounced in SNSPDs, contrary to the situation for SPADs. While some publications have reported afterpulsing effects, such as those described in [30], they assumed that these afterpulses were not caused by the SNSPD itself but by the amplifiers used in the signal processing chain. Afterpulsing in SPADs is strongly related to the dead time of the detector. If long dead times can be accepted, the APP can be reduced to a value that meets the application's needs.

5.2.2.3 Timing jitter

While both single-photon detector types, SNSPDs and SPADs, can achieve very low timing jitter, the timing jitter of SNSPDs can be even lower than that of SPADs. Commercially available counting modules containing SPADs typically achieve timing resolutions in the range from 250 ps [17] to 1 ns [15], with dead times of 22 and 45 ns, respectively. The commercially available SNSPD system ID281 from IDQ achieves a timing jitter of ⩽30 to 60 ps [23] at FWHM and a dead time (or more precisely a full recovery time) of 60 ns.

In [31], B. Korzh *et al* presented a geometrically optimal and material-optimal SNSPD that achieved a timing jitter of only 2.6 ps for visible wavelengths and 4.3 ps at a wavelength of 1550 nm. They also showed that the intrinsic part of the jitter depends on the photon energy. Photons with longer wavelengths and therefore less energy introduce more timing jitter.

Silicon SPADs can achieve timing jitters of the same order of magnitude, as shown in [32], which presented a SPAD integrated in a 65 nm CMOS technology that exibited a timing jitter at FWHM of 7.8 ps at 410 nm. This jitter included the timing jitter of the quenching circuit, which was approximately 4 ps at FWHM.

SPADs capable of detection at 1550 nm typically have increased timing jitter. In [33], a timing jitter of 90 ps at FWHM at 1550 nm was presented, which achieved a PDE of ~30%. An even lower timing jitter of 52 ps at FWHM was presented in [29] for an SPAD using negative feedback. This SPAD used an integrated feedback resistor for quenching and also achieved low dark count rates if cooled to low temperatures. At −110 °C, the dark count rate reached values of around a few cps.

5.2.2.4 Dead time

Both SNSPDs as well as SPADs can reach dead times in the low-nanosecond regime. The advantage of SPADs is that if they are integrated with their quenching circuits, the parasitics are quite low, which theoretically allows fast quenching. In practice, dead times considerably below 10 ns typically result in a large afterpulsing probability, since the lifetime of most traps (i.e. the main reason for afterpulsing in SPADs) is in the range of several ns.

While afterpulsing is not pronounced in SNSPDs, one limiting factor for the dead time in these devices is the difficulty of integrating the amplifiers and the reset circuitry into the same chip as the SNSPD, due to its operation at cryogenic temperatures. The additional signal delay and parasitics are some of the limiting factors for the dead time in SNSPDs.

In [34], a 16-pixel SNSPD array was presented which had a maximum count rate of 1.56 GHz. This might lead to the assumption of a sub-ns dead time. However, the 1.56 GHz represented the combined signals from 16 single pixels, each of which had a dead time of ~4.1 ns. Dead times potentially shorter than 1 ns are possible with the implementation described in [35], which was an SNSPD with a length of only 1 μm. Due to its short length, a decay time (i.e. the transition time of the leading edge of the detection pulse) of 120 ps and a recovery time (i.e. the transition time of the trailing edge of the detection pulse) of ~510 ps was possible. The authors therefore claimed that pulse rates of more than 1 GHz should be feasible.

Similar dead times are achievable in SPADs. A dead time of ~3.5 ns was achieved in a silicon SPAD receiver described in [36]. In [37], a dead time of only 1.93 ns was presented for an InGaAs SPAD.

5.2.2.5 Pixel arrays

In the construction of pixel arrays, SPADs have a clear advantage compared to SNSPDs. The requirement for them to be operated at cryogenic temperatures of ~2 K and even lower complicates the integration of the readout electronics into the same chip as the SNSPD. In contrast, SPADs can be integrated into CMOS, which simplifies the construction of larger arrays. In monolithic solutions, in which the SPADs are on the same chip as the quench and readout electronics, the fill-factor is typically limited. The fill factor can be improved using microlens arrays [38, 39]. An alternative solution is 3D integration, in which the SPADs are on a different chip from the quenching and readout electronics [40, 41].

Table 5.1. Comparison of the key parameters of SPADs and SNSPDs

Parameter	SPAD	SNSPD
PDE @visible	75[2]% [17]	~85%[3] [44]
PDE @1550 nm	36% [25]	98% [24]
Dead time	1.93 ns [37]	4.1 ns [34]
DCR	10 cps[4] [15]/7 cps[5] [29]	0.25 cph[6] [26]/0.5 cps[7] [27]
APP	~0%[8]	~0%
Timing jitter	2.6 ps[9]/4.3 ps[10] [31]	7.8 ps[11] [32]/52 ps[12]. [29]
Pixel array size	1024 × 1024 [42]	32 × 32 [43]
Cryogenic operation required	no	yes

While silicon SPAD arrays reaching 1 megapixel have already been presented [42], the maximum number of pixels is still limited for SNSPDs. Most SNSPD arrays only have a very limited number of pixels, such as the 16-pixel array described in [34]. In [43], a kilopixel array of SNSPDs was presented. However, even though its pixel count was much lower than the pixel count of SPAD arrays, this SNSPD array had to perform row and column multiplexing, which prevented multi-photon detection.

5.2.2.6 Summary

Finally, we offer a comparison of the key parameters of the best available SPADs with the best available SNSPDs in table 5.1. Please be aware that most of these parameters have a trade-off relation with each other. For example, if a high PDE is required, the lowest possible DCR is typically unachievable. Another example would be the minimum achievable dead time versus detector noise, i.e. mainly the DCR and the APP.

5.3 Quantum key distribution

QKD is a physically secure way to share a secret key that can be used to encrypt data that is shared over a (possibly public) channel. The length of this key in relation to

[2] @650 nm
[3] @630 nm
[4] Sensitive in the visible range.
[5] Sensitive at 1550 nm and cooled to $-100\ °C$.
[6] Sensitive in the UV range, cph correponds to counts per hour.
[7] Sensitive at 1550 nm.
[8] For SPADs, the APP typically depends strongly on the dead time. For sufficiently long dead times, the APP can be reduced as close to 0% as necessary.
[9] @532 nm
[10] @1550 nm
[11] @410 nm
[12] @1550 nm

the length of the data defines the level of security. If the key has the same length as the data, or is longer, it can be shown that this encryption cannot be cracked [45].

Many of the currently used encryption schemes rely on Rivest–Shamir–Adleman (RSA) encryption, which ensures a high level of security if the key length is sufficiently long and if only classical computers are available. However, with emerging quantum computers that incorporate more and more qubits, it is to be expected that this type of encryption scheme will be cracked in the near future using the Shor algorithm [46].

While considerable efforts have been made in order to develop new classical encryption techniques that are not affected by the Shor algorithm [46], it is not guaranteed that no other quantum algorithms can be found that will render these new approaches useless again.

In contrast to classical approaches, QKD relies on the fact that a (long) key is shared in physically secure manner between two parties and that this key can be used to encrypt data. Sharing such a typically random key, is, of course, also possible by physically transporting it from point A to point B, i.e. by storing it on a Universal Serial Bus (USB) key and transporting it in a suitcase. However, in this case, you need to trust the person who is transporting the key. Ideally, the key should have the same length as the message, and an encryption method, such as a one-time pad, should be used. Moreover, as will be shown later in this section, each secure key should be used only once. However, it is, of course, also possible to share a key for classical encryption approaches, such as the Advanced Encryption Standard (AES) [46]. Quantum key distribution mainly allows you to share a secret key with a second party. This mechanism is physically secure, meaning that the (currently valid) laws of physics guarantee that if someone is eavesdropping on the transmission of the key, you will be able to detect this and to discard this part of the key [47].

In QKD, it is common to call the transmitting station *Alice* and the receiving station *Bob* (presumably originating from the transmission 'from A to B'), while the eavesdropper is called *Eve* [47]. We will use this naming convention as well.

In this section, we will summarize the basic aspects of QKD in order to enable an understanding of the principles. We will discuss the one-time pad (as a basic encryption scheme) and the two main classes of QKD: one that uses single photons and the other that uses entangled photon pairs. Additionally, we will take a closer look at quantum random number generators (QRNGs), since they are an essential part of QKD and because many QRNG implementations contain SPADs. Given these basic aspects, it will be possible to better understand the requirements for the single-photon detectors used in this type of application. These requirements are discussed at the end of this section.

We highly recommend an excellent review article by Gisin *et al* [47], which contains extensive details of QKD for the interested reader.

5.3.1 One-time pad

The main characteristic of one-time pad encryption is that it uses a secret key that is only known by the two parties that want to share the encrypted data. As the name already suggests, this key may only be used once. Assuming that the key has the same length as the data to be transmitted, one simple way of encrypting the data is by applying an XOR operation that accepts the data as one input and the key as the second input. The data can be decrypted again by applying the XOR operation to the encrypted data and the key, as depicted in the basic block diagram in figure 5.9. Assuming a perfectly random key, the data cannot be extracted without knowing the key [45].

But why can the key only be used once? We assume you are using the XOR operation to encrypt your data. If you encrypt two data streams, *Data1* and *Data2*, with the same key *Key1*, the encrypted output streams are then:

$$Out1 = Data1 \oplus Key1, \tag{5.4}$$

$$Out2 = Data2 \oplus Key1. \tag{5.5}$$

If these two output streams are combined by an XOR operation as well, one gets

$$Out1 \oplus Out2 = (Data1 \oplus Key1) \oplus (Data2 \oplus Key1) = Data1 \oplus Data2, \tag{5.6}$$

i.e. the result is the XOR operation applied to both data streams. The random part of the key is no longer present in this output, which (may) allow an eavesdropper to extract useful data. In [48], a very intuitive method is used to show this more graphically, by encrypting two images with the same key. Using the same method, figure 5.10 shows the encryption of two black-and-white images of the word 'QKD' and a photon symbol that was also used in the previous section. A white pixel in the image corresponds to a digital '0' and a black pixel to a digital '1'. If the XOR operation of both encrypted images is performed, the original data of both images becomes recognizable again.

5.3.2 BB84 protocol

The BB84 protocol is a QKD protocol published by Charles Bennet and Gilles Brassard in 1984 [49]. It is a 'prepare and measure' protocol, meaning that the sender, Alice, prepares a quantum state, which is later measured by the receiver, Bob. The BB84 protocol is a very common QKD protocol, and does not require any kind of entanglement. Figure 5.11 depicts a very basic block diagram showing the main building blocks required for this protocol, assuming that polarization encoding with two different bases is used. Please note that many other encoding schemes exist for QKD [47], however, for an understanding of the basic concept, they do not make

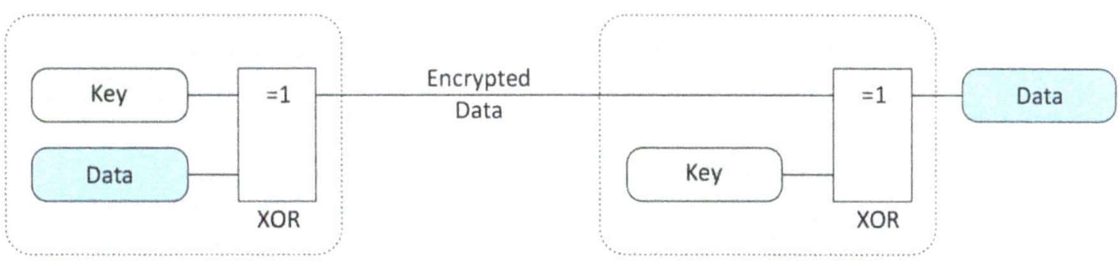

Figure 5.9. Basic block diagram of one-time pad encryption and decryption.

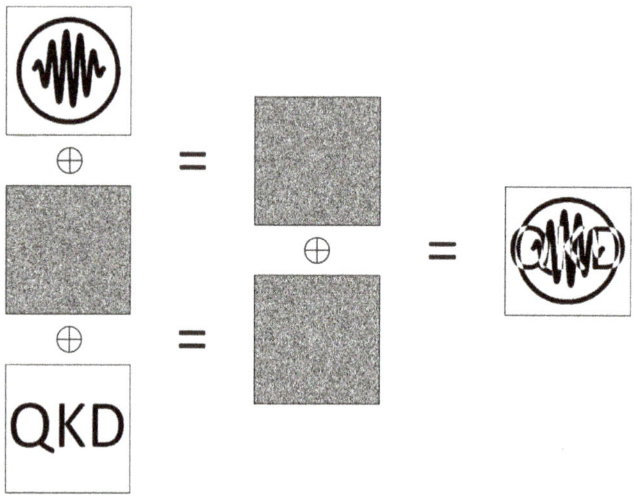

Figure 5.10. Graphical example after [48] of the consequence if a one-time pad is used twice and if both encrypted messages are available. A white pixel corresponds to a digital '0,' and a black pixel corresponds to a digital '1.' The left part shows (from top to bottom) the first image to be encoded, the key, and the second image; the central part shows the encrypted images, both encrypted with the same key; and the right part shows the XOR operation applied to both encrypted images.

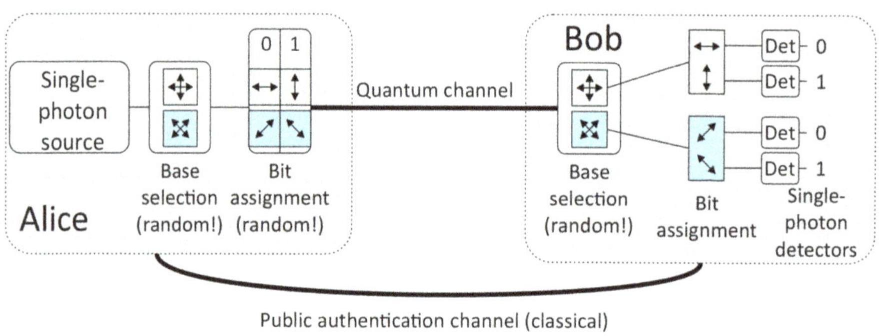

Figure 5.11. Basic block diagram of the BB84 QKD protocol.

any difference. We will therefore focus only on polarization encoding in our discussion of QKD protocols.

The transmitting station, Alice, uses a single-photon source to generate a stream of single photons. It has been shown that instead of a real single-photon source, weak laser pulses can also be used. In that case, depending on the probability that more than one photon is present in a pulse, parts of the resulting key will be lost (i.e. a shorter key is derived from the received key—a process called privacy amplification) in order to ensure that the message cannot be cracked. This is necessary, since in such a case, single bits of the raw key might be known by the eavesdropper Eve, because the protocol is no longer completely secure for pulses containing more than one photon [47].

The next building block is the base selection. In this case, either the horizontal/vertical (H/V) base or the diagonal base is selected. It is of utmost importance that this base selection cannot be predicted by the eavesdropper Eve. Therefore, random number generators that are capable of generating true random numbers are typically used for base selection. SPADs can be utilized in the design of very compact and fast quantum random number generators (QRNGs) [50]. Due to their importance for QKD and since SPADs can be used to build them, we discuss this type of QRNG in more detail in section 5.3.4.

After the base selection, a random bit is assigned using the chosen base and the photon's polarization is set correspondingly. In practical realizations, the base selection and the bit assignment are typically completed in one step by randomly choosing one of the four possible polarization states of the two bases.

The photon is then transmitted over the quantum channel to the receiving station, Bob. This channel can be, for example, an optical fiber or a free-space transmission [5].

Bob also selects a random base (independently from the base selection of Alice!) and, using a polarizing beam splitter, detects which polarization the photon has. Assuming an ideal system, Bob is only guaranteed to detect the polarization that Alice was using for encoding if both Alice and Bob have chosen the same base. If this is not the case, e.g. if Alice has encoded a horizontal photon (state '0' in the H/V base) and Bob has chosen the diagonal base, there is a 50% chance that Bob will detect state '0' (the first diagonal state) and a 50% chance that he will detect state '1.' This fact is crucial for the operation of this QKD protocol. In order to explain this in more detail, we need to take a closer look at a simple transmission example.

The two polarizing beam splitters in the H/V base and in the diagonal base work as shown in figure 5.12. Please note that a polarizing beam splitter for diagonal polarization can, for example, be realized by rotating the polarization of the incoming photon by 45° and then sending it through a standard H/V polarizing beam splitter.

If a photon with horizontal or vertical polarization is present at the input of the H/V beam splitter, it does not change its polarization state when it passes through the splitter, and each photon always exits at the corresponding output (i.e. the horizontally polarized photon exits at the 'horizontal' output and the vertically polarized photon exits at the 'vertical' output).

This situation changes if one of the two diagonal polarization states is present at the input of the H/V beam splitter. In this case, there is a 50% chance that the photon exits at the horizontal or the vertical output. Additionally, when the photon exits the polarizing H/V beam splitter, the photon is no longer diagonally polarized, but horizontally or vertically polarized, corresponding to the output at which it exits.

The same is valid for the diagonal polarizing beam splitter. The polarization states of photons with the two possible diagonal polarization states at the input are not changed when these photons pass through this beam splitter, and the photons exit through the corresponding outputs. Horizontally or vertically polarized photons have a 50% chance of exiting the first or the second diagonal polarized output, and the polarization of these photons changes to the corresponding diagonal polarization state.

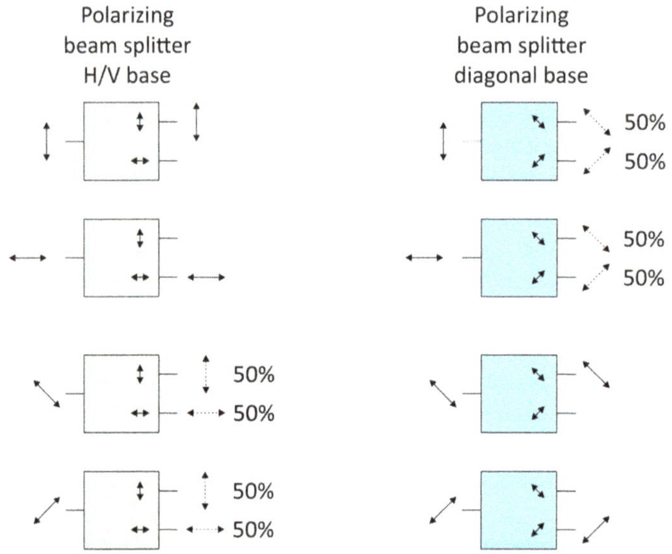

Figure 5.12. Transmission characteristics of an H/V beam splitter and a diagonal polarization beam splitter for horizontal, vertical, and two diagonal polarizations at their inputs.

This property of the polarization state, i.e. that it only can be measured if the correct base is selected, is crucial for the QKD principle. If the wrong base is chosen, first, there is a chance that the wrong bit will be measured relative to the polarization, and second, the polarization of the photon is changed when it passes through the beam splitter [47].

Please note that the non-cloning theorem states that it is not possible to (perfectly) copy an unknown quantum state [47]. It is therefore impossible to create one or several copies of the incoming photon, in order to send these copies through several different beam splitters. In a man-in-the-middle attack, the single photon that was transmitted is destroyed by the eavesdropper, who then has to prepare a new photon in the state he believes the destroyed photon was in.

Let us start with the transmission example in figure 5.13. For the first bit, Alice sends a photon in the randomly chosen H/V base. The random bit to be sent is '1,' therefore, a vertically polarized photon is sent over the quantum channel. For the first bit, the random choice of bases results in the diagonal base for Bob. Since Bob measures in the diagonal base but the photon is vertically polarized, there is a 50% chance that Bob measures a '0' or a '1.' In this first bit, Bob measures '1' by chance, which corresponds to the bit Alice was submitting. Assuming an ideal system, the probability of receiving the correct bit is 50% in this case.

For the second bit, both Alice and Bob choose the same base. Therefore, Bob measures the same polarization that is transmitted by Alice. Alice transmits a '1,' which corresponds to a vertical polarization in this example in this base. Bob detects the vertical polarization, and therefore also the bit '1.' Assuming an ideal system, the probability of receiving the correct bit is 100% in this case.

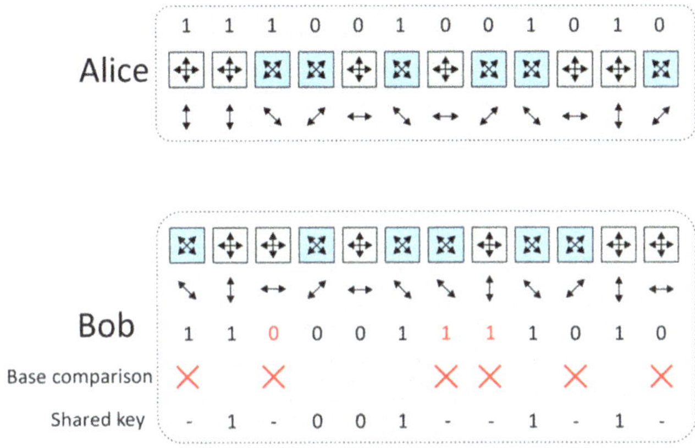

Figure 5.13. Transmission example for the BB84 protocol without an eavesdropper.

For the third bit, Alice and Bob randomly choose a different base again; Alice chooses the diagonal base and Bob chooses the H/V base. Alice transmits a '1.' Since Alice and Bob have chosen different bases, there is a 50% chance that Bob receives the correct bit. In this case, Bob is unlucky, and he receives a '0' instead of a '1.'

The remaining bits work in a similar way to that explained for the first three bits. After the transmission of the key, Alice and Bob use a classical channel that even can be public. They exchange information that describes which base they have chosen for each bit, but not which bit they prepared and measured. By exchanging this information, they know for which bits they will have the same result, namely, for those bits where both have chosen the same base. All bits for which Alice and Bob chose different bases are discarded (this process is called 'sifting'). The bits for which both used the same base form the shared key that only Alice and Bob can know [47].

This key can be used in a subsequent transmission to encrypt the data to be transmitted.

How does this situation change if an eavesdropper is present? Let us assume a man-in-the-middle attack. Additionally, we assume that our single-photon source really only transmits one photon per bit. Due to the non-cloning theorem, the eavesdropper Eve needs to detect the photon (and therefore destroy it), and forward another photon to Bob. Eve's task is to extract the key without being detected by Alice and/or Bob. Let us take a closer look at the transmission example in figure 5.14, which is similar to that in figure 5.13, but now additionally includes Eve.

Let us start with the first bit. Remember that the bases of Alice and Bob are chosen randomly. If the choice of base is truly random, Eve cannot predict this choice. Therefore, Eve can also only randomly pick bases. For the first bit, Alice and Eve choose the same base, while Bob chooses a different base. Since Alice and Eve have chosen the same base, Eve can extract the real polarization state of the photon transmitted by Alice and therefore is able to extract the bit that Alice was transmitting and forward a photon with the same polarization state to Bob. Since Bob has chosen the other base, he only has a 50% chance of receiving the correct bit. However, after the transmission of all the bits, when the comparison of the bases

Figure 5.14. Transmission example for the BB84 protocol, including an eavesdropper (a man-in-the-middle attack).

between Alice and Bob is done, this bit will be discarded, since Alice and Bob have chosen different bases.

A very important case is the one present in the second bit. Alice and Bob have randomly chosen the same base, while Eve used the other base. Since Eve uses a different base than Alice, Eve has a 50% chance of detecting a ‘0’ or a ‘1.’ Additionally, Eve will forward a photon in the diagonal base, or, to be more precise, a photon with a right diagonal polarization. Since Bob detects this photon in the H/V base, he has a 50% chance of detecting the same bit as Eve and consequently also a 50% chance of detecting the same bit as Alice. When the bases are compared after the transmission of all bits, Alice and Bob see that they had the same base and consequently they will use the bit for the key. However, in this example for this bit, Bob has received a ‘0,’ while Alice has sent a ‘1.’ Eve’s detection and forwarding process has caused a bit error. Consequently, the so-called quantum bit error rate (QBER) in the key can be utilized to check whether an eavesdropper was present during the transmission or not. If an eavesdropper was present, the key can be discarded.

But how can the QBER be extracted? A simple way to do this is to compare a part of the shared key over the classical channel. Since this classical channel is insecure, this part of the key needs to be discarded afterwards. If the QBER in the part that is compared is greater than a certain level, this shows that a man-in-the-middle attack is ongoing and that the key needs to be discarded [47].

Figure 5.15 shows a table with all the possible base combinations. The number of possible combinations is $2^3 = 8$, and each base combination has the same probability of 1/8, corresponding to 12.5%. Only those combinations for which Alice and Bob have chosen the same base are valid and part of the shared key. Therefore, only these

Figure 5.15. QBER caused by a man-in-the-middle attack.

combinations need to be considered for the derivation of the QBER introduced by Eve. For two of these combinations, Eve has chosen the same base as Alice and Bob and therefore will not introduce bit errors. For the other two valid combinations, Eve has chosen a different base from those chosen by Alice and Bob. For both of these cases, there is a 50% chance that Eve has introduced a bit error. To summarize: in 50% of the valid cases, Eve chose the wrong base, which resulted in a bit error for 50% of the transmissions. Consequently, Eve's man-in-the-middle attack introduced a QBER of 50% of 50%, or 25%.

A realistic QKD transmission system will also suffer from systematic errors that cause a QBER larger than zero, even if no eavesdropper is present. These systematic errors are corrected by means of classical error correction algorithms, which will (slightly) reduce the effective number of bits in the shared key [47]. In order to reliably detect an eavesdropper, these systematic errors need to be kept below the level an eavesdropper would cause. Sources of systematic errors are, for example, imperfect state preparation by Alice, and dark counts and afterpulsing in Bob's detectors [63]. Another imperfection is due to channel loss. If the photons are transmitted over an optical fiber, a loss of ~0.2 dB km^{-1} can be expected if a low-loss fiber is used (e.g. the commercially frequently used SMF-28 [20]). This fiber loss, together with detector noise, limit the maximum length of the fiber for QKD to approximately 200 km for reasonable bit rates [3]. While some publications have described transmission lengths of approximately 400 km or (recently) even more than 500 km [4], the resulting key rates (i.e. the effective bit rates of the secure bits) are very low (i.e. in the range of 6.19×10^{-9} bit per signal pulse). For a fiber length of 1000 km, the loss would reach 200 dB. If 10^9 photons/s were transmitted by Alice, this channel loss alone (without any other imperfections) would result in a photon rate of $\sim 10^{-11}$ photons/s for Bob, which is equivalent to one photon in more than 3000 years.

Amplification of the fiber signal is not possible for QKD, which is why, for longer distances, trusted nodes are currently required after a certain distance to guarantee reasonable key rates. For example, in the 2000 km fiber link from Beijing to Shanghai, 32 trusted nodes are used [5]. One possibility for extending the distance without using trusted nodes would be the implementation of quantum repeaters. However, so far, no working quantum repeater has been presented.

Another possibility for covering larger distances is satellite-based QKD, since it is possible to cover large distances without photon loss in space [5]. In 2017, Liao *et al* published the first results for satellite-based QKD using the satellite Micius [51]. This satellite allowed different QKD protocols to be investigated. If BB84 is used, the satellite itself needs to be considered a trusted node. However, using entangled photons, for example, in the Ekert protocol, which is briefly introduced in the next section, the satellite can be used to allow the exchange of a key between two different ground stations without the necessity of trusting the satellite [52].

5.3.3 Ekert protocol

The basic principle of the Ekert protocol, published by Artur Ekert in 1991 [53], is quite similar to that of BB84. The main difference is the use of an entangled photon source instead of a single-photon source. This entangled photon source is not necessarily placed at Alice's or Bob's station, but can, in principle, be placed anywhere, e.g. in the middle between Alice and Bob. This protocol is secure, even if the eavesdropper Eve controls the entangled photon source, which has important implications, e.g. if this protocol is used in satellite-based QKD. Figure 5.16 depicts a basic block diagram for QKD using the Ekert protocol.

In the Ekert protocol, three bases are typically used, and Alice and Bob share only two of them. The basic protocol is quite similar to that of BB84. Alice and Bob randomly choose their bases for photon detection. Due to the use of an entangled photon source, the detection results of Alice and Bob are correlated. Let us assume that the source is entangled in such a way that both photons always have the same polarization. If Alice and Bob are measuring using the same base, they will detect the same polarization. These cases can be used for the shared key. If Alice and Bob have chosen different bases, their measurement results still follow the correlation statistics of entangled particles. These statistics are different from those of independent particles. In the entangled case, the measurement statistic will violate Bell's inequality [47]. A test for the violation of the Bell inequality therefore shows Alice and Bob that they were receiving both one photon of an entangled photon pair. If an eavesdropper intercepted one or both of these photons, he would destroy the entanglement, which can be detected by Alice and Bob.

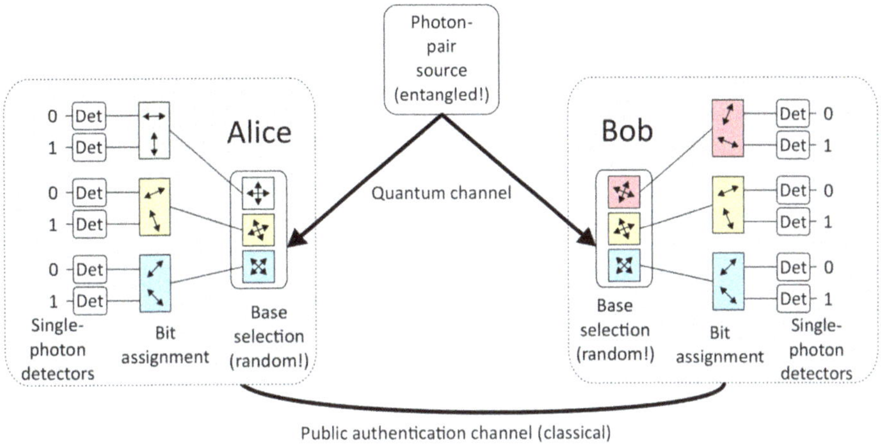

Figure 5.16. Basic block diagram of an Ekert QKD protocol.

This approach is especially interesting for satellite-based QKD. Satellite-based QKD allows keys to be shared between more distant locations than is possible with optical fibers [5]. If the the Ekert protocol is used and the entangled photon source is placed in the satellite, QKD can take place between two ground stations without the requirement for the satellite to be a trusted node. A technological difficulty is that both ground stations, Alice and Bob, need to have simultaneous contact with the satellite. Additionally, the attenuation due to transmission through the atmosphere and beam dispersion needs to be considered twice, since Alice and Bob need to receive photons from the same pair. The Micius satellite was able to distribute entangled photon pairs to two different ground stations separated by more than 1200 km with a simultaneous photon detection rate of approximately 1 Hz [52].

5.3.4 Quantum random number generator

As seen in the previous sections, one of the most critical key components of any QKD system is a random number generator capable of delivering true random numbers. If an eavesdropper is able to predict the choice of the base used for the transmission or the detection of the photons, the transmission is no longer secure, and a man-in-the-middle attack becomes feasible. The physical secure transmission scheme therefore stands or falls on the quality and the purity of the randomness used to select the base. Random number generators are typically grouped into two separate categories. First, pseudo random number generators rely on mathematical algorithms that are typically initialized using a seed. Their deterministic nature prevents their utilization in QKD. The second group are true random number generators, which typically rely on undetermined physical effects. A subgroup of these true random number generators is the quantum random number generators, which benefit from the probabilistic nature of quantum effects [50]. In QKD, as well as being truly random, the random number generator needs to be sufficiently fast to provide two new random bits for any transferred bit: one for the key and one for the selection of the base. A very simple approach for generating a sequence of random bits is the utilization of a beam splitter, as depicted in figure 5.17 [54]. Assuming a split ratio of 50%, a random sequence can e.g. be generated by assigning the bit a '0' if the photon appears at the first output and a '1' if it appears at the second output.

Another method for generating quantum random numbers, which has the potential for a high degree of integration, is the utilization of the Poisson statistics of light [50]. It can be shown that measuring and comparing the interarrival times (IATs) of photons can be used in the generation process. Such a method is depicted in figure 5.18. The sequential IATs are ordered in non-overlapping pairs and the bits corresponding to each pair are derived e.g. according to:

$$\text{Output} = \begin{cases} 1 & \text{if } t_{2i-1} > t_{2i}, \\ 0 & \text{if } t_{2i-1} < t_{2i}, \\ \text{Abandoned} & \text{if } t_{2i-1} = t_{2i} \end{cases} \tag{5.7}$$

$$\text{for } i = 1,2,3,\ldots \tag{5.8}$$

If the first IAT of a pair is longer than the second one, the bit corresponding to this pair is a logical '1'; if it is shorter, then the corresponding bit is a '0.' If the two IATs of a pair have the same length (within the measurement accuracy), the corresponding bit is discarded.

Why is it important that the pairs are non-overlapping? If they were overlapping, two consecutive bits would not be statistically independent. Let us take a look at the example in figure 5.19. In this example, the second IAT is much longer than the first one. This would result in a logical '0' for this pair. If a pair that overlaps with the first one were taken, namely the pair formed by t_2 and t_3, the probability would be

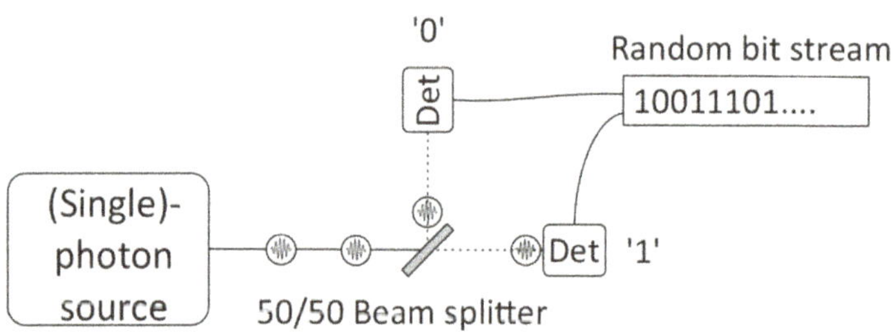

Figure 5.17. Simple quantum random number generator (QRNG) using a beam splitter after [54].

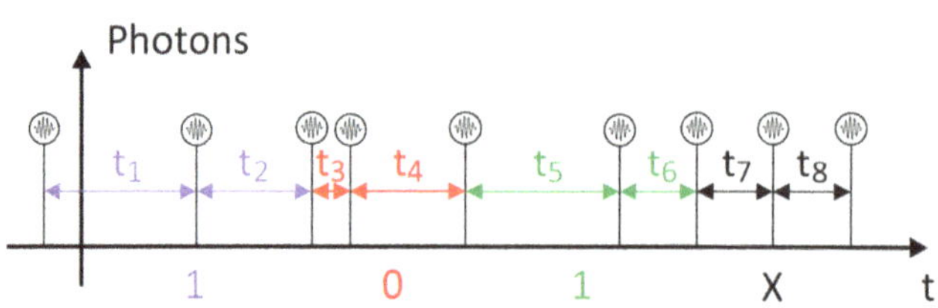

Figure 5.18. Quantum random number generator (QRNG) utilizing the Poisson statistics of light.

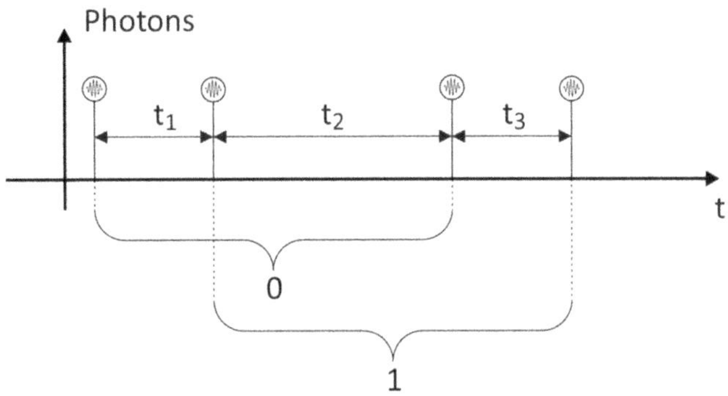

Figure 5.19. Example of statistical dependence if overlapping pairs are used.

higher that the next bit is a '1,' since the second IAT was a long one. Consequently, with overlapping pairs, the probability of alternating bits would be higher than the probability that two consecutive bits would have the same value. Such a statistical dependency could be utilized by an eavesdropper to predict the next base to be used by Alice and/or Bob.

This version of a QRNG is especially interesting because it allows a high degree of integration. The detection of the train of photons can be accomplished by an SPAD. The use of an SPAD in a CMOS process allows the integration of the quenching circuitry as well as a circuit for comparison of the time pairs. In [50, 55] a light-emitting diode (LED) was also integrated onto the same chip as the SPAD. A cross section of this structure is shown in figure 5.20.

Figure 5.21 shows a chip photo of a fully integrated QRNG.

While silicon, as an indirect semiconductor, is a very inefficient light emitter, and normally not used for the manufacture of LEDs, the required light intensity in a QRNG is so low that this poor efficiency is not a problem. To illustrate the order of magnitude involved: for light with a wavelength of 1 μm (i.e. with a photon energy of $\sim 2^{-19}$J per photon) and a photon rate of 10^8 photons/s, the corresponding optical power is approximately 20 pW. The silicon LED is biased in reverse operation in the avalanche regime. According to Bude *et al* [57], luminescence in silicon is driven by hot carriers and dominated by intraband phonon-assisted relaxation processes in the conduction band. In [50, 55, 56] the SPAD is ring-shaped and surrounds the LED; it therefore allows very efficient coupling of the photons from the LED into the SPAD. This method therefore allows a single-chip solution for a QRNG. The QRNG in [50] achieves a bit rate of 1 Mb s^{-1} and passes all the randomness tests from the statistical test suite of the National Institute of Standards and Technology (NIST, USA) [58], as well as ENT [59], and DIEHARD [60], which are other test suites for statistical randomness.

Bisadi *et al* investigated the utilization of silicon nanocrystal LEDs for the generation of the photons in [61] using a similar QRNG. They used an external SPAD instead of an integrated one and achieved a bit rate of 0.6 Mbps.

The maximum achievable bit rate in an approach that uses the Poisson distribution of photons is limited by the Poisson distribution itself and by the

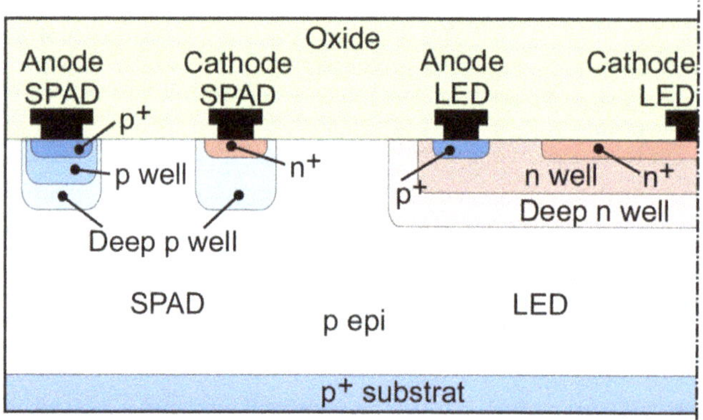

Figure 5.20. An LED integrated together with an SPAD on the same CMOS chip as the source and detector for a quantum random number generator [50].

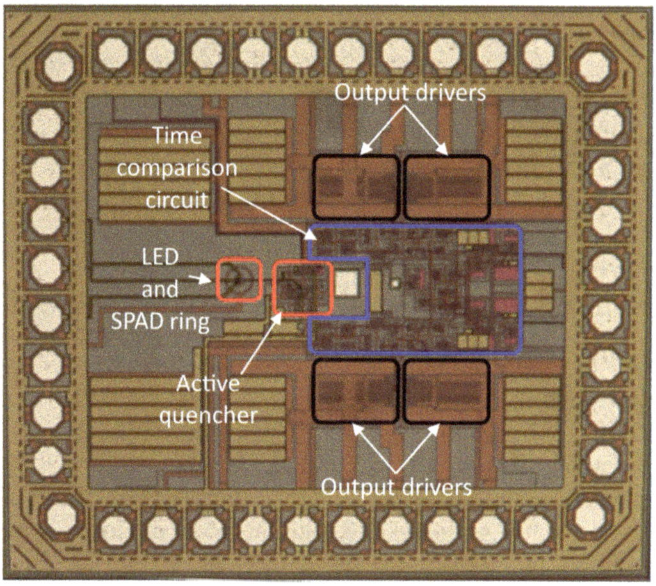

Figure 5.21. Fully integrated QRNG with SPAD and active quenching circuit, LED, time comparison, decision logic, and output drivers [50, 56].

dead time of the single-photon detector [62]. For an average photon rate, a Poisson distribution means that there is a probability of long pauses without photons. The probability of occurrence of these long pauses decreases with increasing photon rates. However, if the photon rate is too high, saturation effects in the single-photon detector degrade the randomness of the generated sequence. This can easily be understood if we consider the extreme case, in which a single-photon detector always triggers after the dead time. In this case, the output signal is a pulse train with a quasi-constant interarrival time, which no longer contains any randomness.

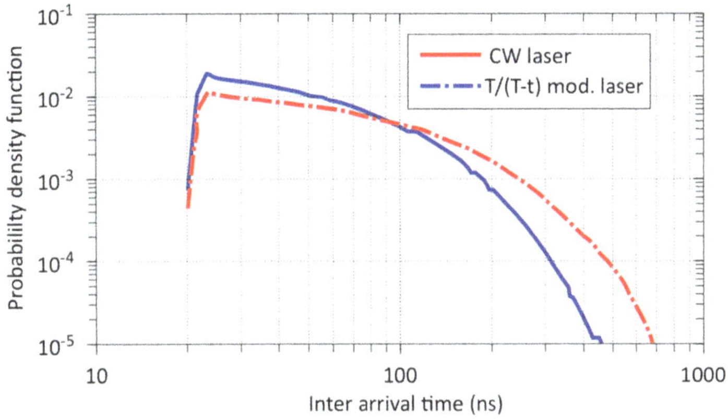

Figure 5.22. Probability density function of a CW laser and a T/(T-t) modulated laser, both adjusted to the same average photon rate. The modulation frequency was 40 MHz. Reproduced from [62] with permission from SPIE.

One way to increase the bitrate of a single QRNG is to use a specially modulated light source. It can be shown that if a laser is periodically modulated with a T/(T-t) pulse shape, the probability density function of the interarrival times can be squeezed, compared to that of a standard Poisson distribution, while keeping the same photon rate, as shown in [62].

Figure 5.22 shows the measured probability density function of the interarrival time of a modulated laser compared to that of a continuous-wave laser, both at the same photon rate. It is clear that the modulated distribution is compressed into shorter times. In [62], a post-processing-free bit rate of 5 Mbps was reached, while passing all the tests of NIST's statistical test suite.

5.3.5 Requirements for single-photon detectors in QKD

Which are the most critical key parameters of a single-photon detector used for QKD? Of course, the photon detection efficiency is important. However, it is not as critical to get as close to 100% as it is for quantum simulation, quantum computing, or quantum super-resolution microscopy, for example. A PDE of less than 100% is equivalent to additional loss. As we know, for fiber-based QKD, tens of decibels of loss are caused by the fiber alone for a reasonable fiber length; an imperfect PDE will (slightly) reduce the achievable length, but will not be a showstopper for the technology.

A low-loss single-mode fiber has an attentuation of ~0.2 dB km^{-1} [20]. Therefore, if the detector has a loss of 2 dB (i.e. a PDE of 63%), this corresponds to a 10 km loss for this fiber.

For state-of-the-art counting modules with silicon SPADs, one can expect peak PDEs in the range of 70% [15]. This value corresponds to an additional loss of 1.55 dB. However, silicon SPADs do not operate in the telecom wavelength range of 1550 nm, where these low-loss fibers are available. A single-mode fiber designed for red light (i.e. at around 630 nm) has a loss that is considerably higher. The S630-HP single-mode fiber, which has a wavelength range of 630–860 nm has an attenuation of $\leqslant$10 dB km^{-1} at 630 nm [21]. This high loss is unacceptable for long-range fiber-based QKD.

Unfortunately, as discussed in section 5.2.2, SPADs in the telecom wavelength range have considerably lower PDEs than those of silicon SPADs. If we assume a PDE of 36%, as in [25], this equates to an additional loss of −4.44 dB caused by the single-photon detector, corresponding to the loss caused by a ~20 km fiber. However, this may still be acceptable for many applications.

If the range of the link needs to be maximized, one currently has to take advantage of the higher PDE of SNSPDs, which recently reached system detection probabilities of 98% [24]. However, this comes at considerable additional cost and the SNSPD needs to be operated at cryogenic temperatures. It can be assumed that for commercial QKD systems, the additional loss of the SPAD will be accepted.

A photodetector's noise, i.e. mainly its dark count rate and its afterpulsing probability, is more critical than its loss. The dark counts and the afterpulses cause quantum bit errors (QBERs) that, if too numerous, will prevent Alice and Bob from reliably detecting an eavesdropper [63].

With increasing link length and therefore increasing attenuation introduced by the optical fiber, the photon rate at the detector decreases. Therefore, if we assume constant detector noise, the ratio between successfully detected photons and detector noise will decrease. At a certain attenuation, and corresponding to this, at a certain fiber length, the QBER introduced by detector noise will become too big [63].

While silicon SPADs as well as SNSPDs can reach dark count rates of even less than one count per second, the situation is different for SPADs in the telecom wavelength range. In this range, the dark count rate is typically considerably higher, which necessitates cooling in order to reduce this rate.

In summary, for QKD, detector noise is more critical for the operational principle than the PDE. While a high PDE is still beneficial, whether it is 90% or 99% does not make a huge difference. We will see later, however, that this can make a huge difference for other applications.

5.4 Photonic quantum simulation

A quantum simulator is, by definition, a controllable system that allows the emulation or simulation of other quantum systems [64]. There are many different types of quantum simulator. We will briefly discuss two prominent examples, the quantum walk and boson sampling. Both can be implemented in passive waveguide structures. In 2020, Zhong *et al* claimed to have reached quantum supremacy with a quantum simulator that performed boson sampling [1].

5.4.1 Quantum walk

The quantum walk is, in one of its simplest forms, a quantum version of the classical Galton board as shown in figure 5.23. For an increasing number of stages and of balls passing through the Galton board, the distribution of balls at the end converges towards a normal distribution. For a finite number of stages, the distribution of balls follows a binomial distribution [65].

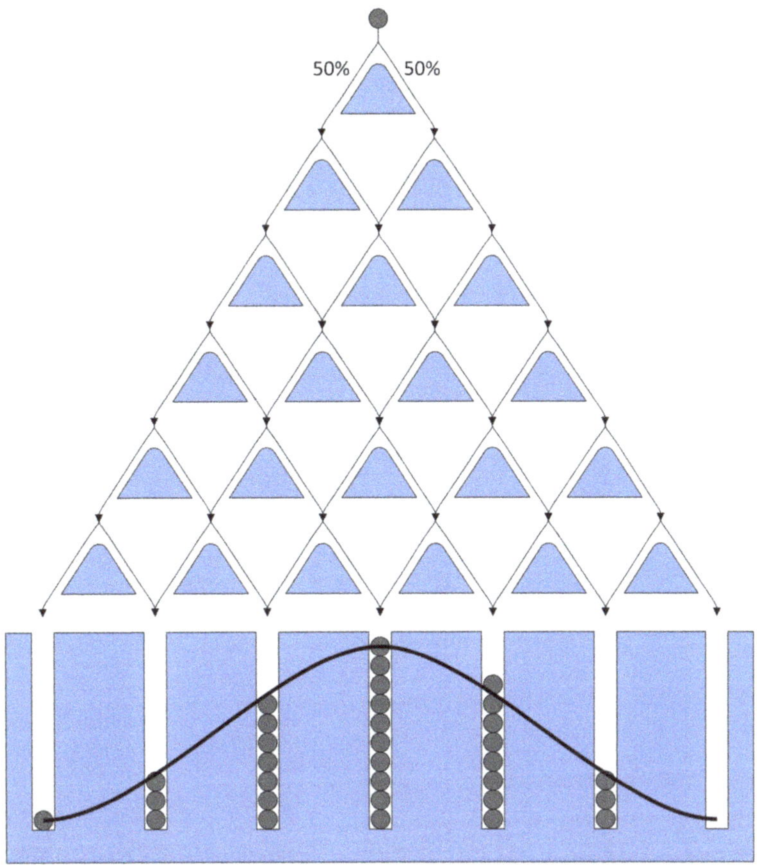

Figure 5.23. Classical Galton board with six stages. The distribution of balls at the bottom follows a binomial distribution for a large number of balls. As the number of stages tends toward ∞, the distribution converges toward a normal distribution.

A simple version of the quantum walk is built very similarly. Beam splitters are used in order to split the propagation path of the photon in two; both paths have the same probability. For a practical implementation, integrated waveguides can be used, for example, as depicted in figure 5.24.

In this example, the photon can either be injected into the left input or the right input, both at the top of the structure. Additionally, superpositions of these two states are possible. Since the photon will interfere with itself, the outcome of the experiment can be very different from that of the classical experiment, and depends strongly on the state of the input photon, as shown in figure 5.25, for a simple quantum walk with 100 steps in a similar but extended structure to that of figure 5.24, according to [66]. If the photon is injected into the left input, the distribution is concentrated on the left-hand side in this example, while it is concentrated on the right-hand side if the photon is injected into the right input. Please note, if the photons are detected after a single splitter of this structure, the

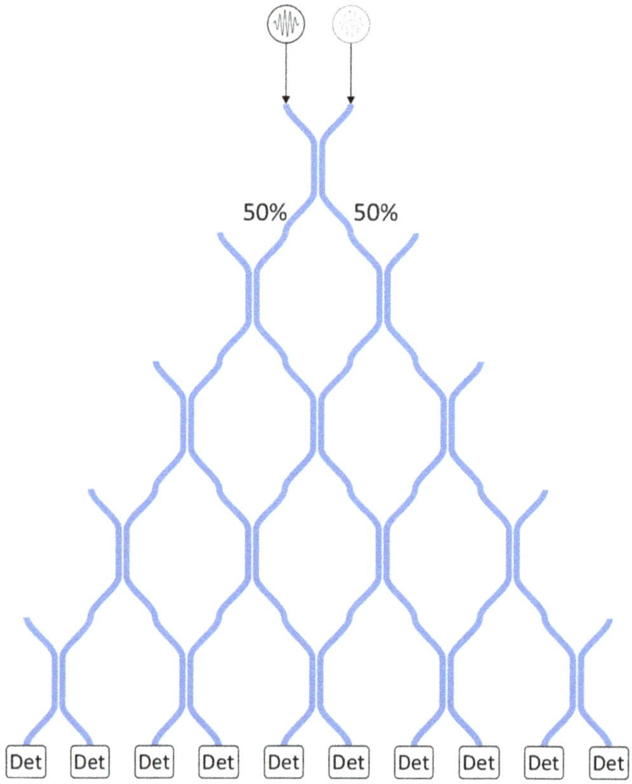

Figure 5.24. Quantum walk using waveguide beam splitters. The distribution detected at the output strongly depends on the state of the photon at the input. The photon at the input can be inserted into the left or the right input. Additionally, superpositions of these two cases are possible.

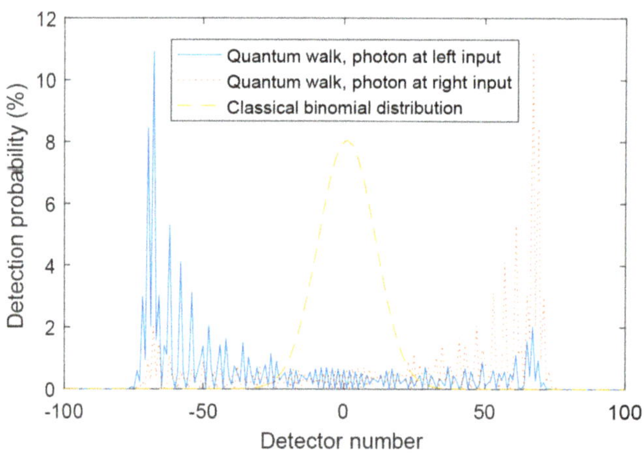

Figure 5.25. Example of the outcome of a quantum walk for a structure as shown in figure 5.24 but with 100 steps, for a photon injected into the left input and a photon injected into the right input, after [66]. Additionally, the classical binomial distribution is shown.

detection probability is 50% for both sides and is independent of the input at which the photon was inserted (the left or the right one).

A discrete implementation of a photonic quantum walk is presented in [67]. Broome *et al* were also able to tune the outcome of the experiment from the quantum outcome to a classical outcome, by tuning the decoherence.

In [68] and [69], quantum walk realizations are presented that used waveguides integrated into glass by femtosecond laser writing. The use of this kind of integration allows the quantum walks to be scaled up to larger number of stages more efficiently than with discrete setups. In both works, two polarization-entangled photons are used instead of a single photon. Due to the symmetry of the entanglement, different particle statistics, such as fermion and boson statistics, can be emulated by the quantum walk experiment. The structure is similar to that of figure 5.24, but the structure described in [68] includes more stages and additional phase shifters in one arm of each splitter, which makes this implementation more flexible. Using two polarization-entangled photons and the ability to adjust the phases in the quantum walk setup, the authors of [68] observed the onset of Anderson localization, which is a quantum effect that causes the diffusion of quantum particles to stop in a disordered potential.

According to [70], universal quantum computing could even be feasible using the quantum walk approach, based on the use of more complex graphs than the simple cascaded beam splitter structure.

5.4.2 Boson sampling

In boson sampling, N photons are inserted into an interferometer with M spatial modes, where M needs to be larger than N. The task of boson sampling is to derive a possible sample of photons at the output of the interferometer (i.e. where output photons arrive). This task is very hard to complete with classical computers. While boson sampling seems to be infeasible for universal quantum computing, it is very interesting, because due to its relatively simple structure, it has the potential to achieve quantum supremacy relatively early. Additionally, boson sampling is perfectly suited for photonic systems, since interferometers can be implemented efficiently [71]. Figure 5.26 depicts a boson sampling setup with M=8 modes, in which N=4 photons are injected.

According to [71], boson sampling corresponds to deriving the permanent of an $N \times N$ matrix. Deriving the permanent of a matrix is quite similar to deriving the determinant of a matrix, just without the alternating sign in the summing process. For an $N \times N$ matrix with coefficients $a_{i,j}$, the permanent is defined as [72]:

$$perm(A) = \sum_{\sigma \in S_N} \prod_{i=1}^{N} a_{i,\sigma(n)}, \tag{5.9}$$

where S_N is the symmetric group containing all possible permutations of the numbers from 1 to N. For an $N \times N$ matrix, $N!$ summands need to be derived and added subsequently. Even for a photon number of $N = 20$, this task is almost

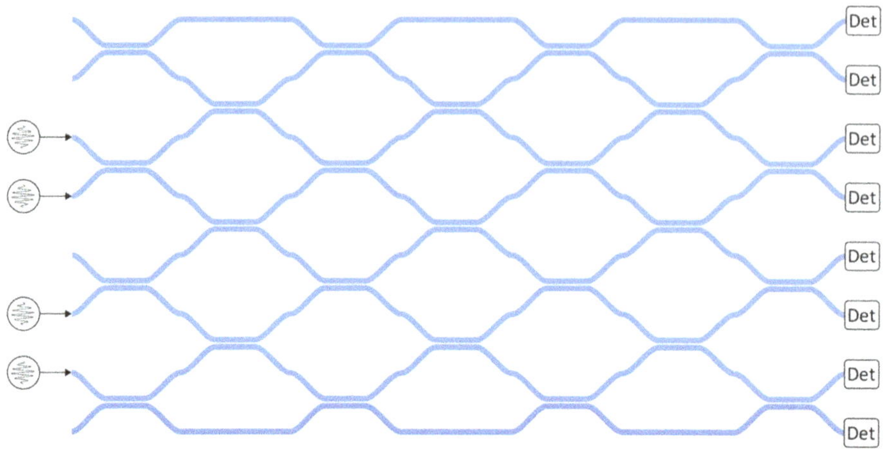

Figure 5.26. Boson sampling with eight modes using waveguide splitters with four injected photons.

impossible to finish on a classical computer[13]. In photonic quantum simulation, the on-demand generation of 20 single photons is challenging. However, the construction of an interferometer to send them through is technologically feasible. In 2017, Wang *et al* demonstrated boson sampling with three, four, and five photons in an interferometer with nine spatial modes [71]. A modified approach using 50 single-mode squeezed states, which were sent through a 100-mode interferometer to perform Gaussian boson sampling, was the method first used to reach quantum supremacy using a photonic quantum system [1]. The authors claimed to be able to sample $\sim 10^{14}$ times faster than by simulating this scenario in a state-of-the-art supercomputer. While the authors of [71] were still able to use silicon SPADs with a PDE of ~32%, for the experiment that reached quantum supremacy ([1]), 100 SNSPDs were used with an average PDE of 81%. A very promising scalable integrated photonic chip that performed Gaussian boson sampling was presented in [73]. While it does not yet allow the same number of modes as [1], the high level of integration promises a larger number of squeezed states and modes in the future. As well as Gaussian boson sampling, this chip also simulates vibronic spectra and solves graph similarity problems.

5.4.3 Requirements for single-photon detectors in quantum simulation

The requirements for single-photon detectors in quantum simulation strongly depend on the kind of simulation performed. For example, in a simple quantum-walk experiment, only one or two photons will typically be present and it (they) can be found at one or two of the outputs of the setup. Since only a few photons need to be detected in each iteration of the experiment, a high PDP is desirable but not as critical as for a boson sampling experiment, for example. Compared to the other losses in the experiment in a typical state-of-the-art setup, a lower PDE can also be

[13] 20!=2 432 902 008 176 640 000.

acceptable. Therefore, even at 1550 nm, SPADs can be a good option for such an experiment. The lower PDE necessitates a larger number of experimental iterations in order to obtain sufficient statistics. Nevertheless, for experiments with only one photon, in particular, an iteration with no detection caused by the limited PDE will not have any other detrimental effect on the result apart from requiring more iterations, since iterations without detection at any output can simply be discarded.

Detector noise, such as dark counts or afterpulsing, can also be critical for quantum-walk experiments. If counts occur at more than one output, this iteration of the experiment can still be discarded, but if only one of the outputs shows a dark count or an afterpulse, this cannot, in some cases, be distinguished from a real detection and might therefore distort the output distribution. This negative effect can be reduced by cooling the SPAD to reduce the DCR [74] and/or by increasing the dead time in order to reduce the APP. However, increasing the dead time also limits the maximum repetition rate of the experiment and therefore increases the time required to perform the experiment.

Another option for reducing the influence of detector noise is to utilize coincidence detection. If a photon-pair source (in which one photon is used as the signal photon, while the other is used as an idler photon to indicate that a photon is present in the experiment and when it is present) or a quasi on-demand single-photon source is used to feed the quantum walk, the timing information of the photon entering the experiment can be used for coincidence detection at the output. If the detector shows a pulse outside the time period in which the photon is expected to arrive at the detector, also considering the timing jitter of the detector, the count was most probably a dark count and can be neglected. However, this coincidence measurement will only reduce the effective dark count rate if the timing jitter is considerably lower than the repetition rate of the experiment.

For boson sampling, the situation is quite similar regarding detector noise. In this case, detector noise should also be reduced to very small levels in order to minimize the distortion of the output distributions. The methods described above can be applied, i.e. reduction of temperature, increase of the dead time, as well as coincidence measurements. However, the requirements for PDE are far more critical in boson sampling than in the quantum walk. In the Gaussian boson sampling[14] experiment described in [1], Zhong *et al* performed an experiment with up to 76 simultaneous detections at the 100 outputs. In order to have a realistic chance of detecting such a large number of photons simultaneously, the PDE needs to be as large as possible. This is technologically challenging, even for SNSPDs, as soon as the channel number increases. In [1], the detection efficiencies of the single channels ranged from 73% up to 92%, also including the coupling losses.

For Gaussian boson sampling, photon-number-resolving detectors are required. A method of achieving photon-number resolution in SNSPDs as well as in SPADs is to utilize arrays of detectors [1]. The transition-edge sensor is another photon-number-resolving detector. However, this sensor needs to be operated at cryogenic

[14] Gaussian boson sampling is a variant of boson sampling in which single photons are not used, but rather single-mode squeezed states of light

temperatures, and it typically has longer dead times than those of SPADs and SNSPDs [75].

5.5 Photonic quantum computing

Great progress has recently been made in the performance of current quantum computers. In 2019, Google claimed to have achieved quantum supremacy, i.e. they claimed that they could solve a problem with their 53-qubit Sycamore quantum computer faster than it would have been possible with any classical supercomputer. They claimed to be able to measure a probability distribution in 200 s that would require more than 10 000 years for a state-of-the-art supercomputer [76]. This quantum computer was not a photonic one, but utilized superconducting qubits.

The main difference between quantum simulation and quantum computing is that quantum computing is universal; in quantum computing, the complexity that can be simulated or derived only depends on the number of qubits, while a quantum simulator is tailored to simulate or emulate a very specific quantum problem or a set of problems. A set of criteria has been developed that need to be fulfilled in order to build a universal quantum computer [77]. In the next part, these criteria will be presented, and we will discuss which of these are the most critical for photonic quantum computing.

5.5.1 Requirements for quantum computers

DiVincenco defined the following five criteria required to build a universal quantum computer [77]:

- *A scalable physical system with well characterized qubits.*
- *The ability to initialize the state of the qubits to a simple fiducial state, such as* $|000\ldots\rangle$
- *Long relevant decoherence times, much longer than the gate operation time*
- *A 'universal' set of quantum gates*
- *A qubit-specific measurement capability*

Additionally, the following two criteria were added, which mainly address the transport of qubits for quantum communications [77]:

- *The ability to interconvert stationary and flying qubits*
- *The ability to faithfully transmit flying qubits between specified locations*

For photonic quantum computing, the scalability as well as the 'universal' set of quantum gates are currently the most challenging parts. For a universal set of quantum gates, two-input gates are also required, which are difficult to build in photonic systems, due to the bosonic nature of photons, which prevents (efficient) interaction between two photons. Solutions exist; however, these solutions require large numbers of on-demand single-photon sources as well as large numbers of detectors. Currently, the on-demand single-photon source is the technologically more challenging part, since SNSPDs allow high photon detection efficiencies. Since two-photon gates are essential and especially challenging in photonic quantum computing, we will discuss a method that only uses linear optical elements and measurements in this section in a bit more detail by inspecting an example of a two-input quantum gate, the CNOT gate. Before this, we briefly explain the qubit, the basic data unit of a quantum computer. We recommend a book by W Scherer [78] to readers who are interested in further details of the mathematics of quantum mechanics.

5.5.2 Qubit

In quantum computing, the basic data unit is not a bit that has only two possible states, namely '0' and '1,' but a qubit, which is defined by two measurable states, e.g. the horizontal and vertical polarizations of a photon. While the measurement outcome in the H/V base is always either a horizontal or a vertical polarization, the qubit itself can have any superposition of these two polarizations. For polarizations in particular, this is quite easily imagined. For a linearly polarized photon, if it is diagonally polarized, measuring the polarization with an H/V polarizing beam splitter will result in either a horizontally or vertically polarized photon. In the case of a diagonally polarized photon, the chance for both options is 50% each. If the polarization of the diagonally polarized photon is rotated so that it is close to a horizontal polarization, the measurement outcome will be horizontally polarized more frequently, and vice versa. The polarization of a photon can, in principle, have an infinite number of different states (i.e. any angle of linear polarization, circular polarization, elliptical polarization, etc). This is the strength of a qubit. In contrast to a classical bit, it can have an infinite number of states. Please note that there are many other possible implementations of qubits besides polarization. However, for the basic principles, the type of qubit is unimportant. Therefore, we will focus on polarization encoding in this chapter.

A general qubit state is defined as [78]:

$$|\Psi\rangle = c_0|0\rangle + c_1|1\rangle \tag{5.10}$$

$$\text{with } c_0^2 + c_1^2 = 1, \tag{5.11}$$

where $|0\rangle$ and $|1\rangle$ are the two measurable base states—in our example above, these would be the horizontal and the vertical polarizations; c_0 and c_1 are complex amplitudes. The squared absolute value of c_0 (c_1) gives the probability of measuring $|0\rangle$ ($|1\rangle$) if the state $|\Psi\rangle$ is measured in the corresponding base.

The Bloch sphere is a geometrical representation of a qubit, as depicted in figure 5.27. The poles of this sphere indicate the base states. In our example above, these would be the horizontal and vertical polarizations. Any point on the surface of this sphere corresponds to a valid state in polarization encoding. Such a general state can be written as follows, using the angles and geometric relations of the Bloch sphere in figure 5.27 [78]:

$$|\Psi\rangle = \sin\left(\frac{\theta}{2}\right)|0\rangle + \cos\left(\frac{\theta}{2}\right)e^{i\phi}|1\rangle \tag{5.12}$$

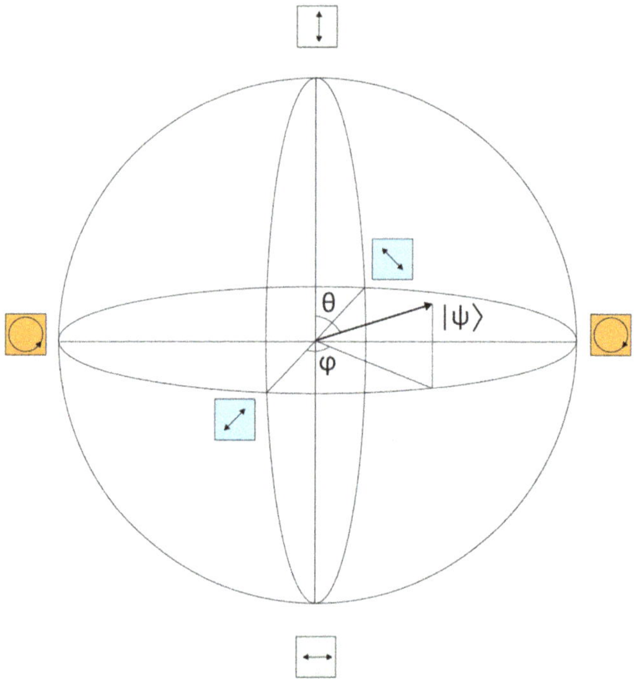

Figure 5.27. Bloch sphere of a polarization encoded qubit, after [78].

For a system composed of two qubits $|\Psi_1\rangle$ and $|\Psi_2\rangle$, the two-qubit state can be obtained by applying the tensor product

$$|\Psi_{12}\rangle = |\Psi_1\rangle \otimes |\Psi_2\rangle. \tag{5.13}$$

The result is then called a product state. All one-qubit states can be combined into product states [78]. Let us assume we have the two following states:

$$|\Psi_1\rangle = \frac{1}{\sqrt{2}}(|0\rangle + |1\rangle),\ |\Psi_2\rangle = \frac{1}{\sqrt{2}}(|0\rangle - |1\rangle); \tag{5.14}$$

the corresponding product state $|\Psi_{12}\rangle = |\Psi_1\rangle \otimes |\Psi_2\rangle$ then equals:

$$|\Psi_{12}\rangle = \frac{1}{2}(|0\rangle \otimes |0\rangle - |0\rangle \otimes |1\rangle + |1\rangle \otimes |0\rangle - |1\rangle \otimes |1\rangle), \tag{5.15}$$

or, in short:

$$|\Psi_{12}\rangle = \frac{1}{2}(|00\rangle - |01\rangle + |10\rangle - |11\rangle). \tag{5.16}$$

For this state, the same steps can be performed in the reverse direction, i.e. this two-qubit state can be separated into two one-qubit states. From a physical point of view, this means that these two one-qubit states can be measured statistically independently of each other.

By using the tensor product repeatedly with one-qubit states, one can generate N-qubit states.

There are, however, states that cannot be separated. These states make quantum physics especially exciting, since they behave differently from our general experience.

An example of such a state is the following:

$$|\Psi_{12}\rangle = \frac{1}{\sqrt{2}}(|01\rangle + |10\rangle). \tag{5.17}$$

This state cannot be separated into two one-qubit states. The physical consequences are remarkable. The state of the first qubit cannot be measured independently of the state of the second qubit. Such states are called entangled states. The specific state above is a state in which the second qubit will always have a different measurement result than the first one. If $|0\rangle$ is measured for the first qubit, the result when measuring the second qubit will always be $|1\rangle$ and vice versa. But remember, the two qubits can, for example, be realized by two photons. These two photons can be far away from each other, and this correlation will still always hold.

A general N-qubit state that also includes states that cannot be separated (i.e. those that are entangled), can be written as [78]:

$$|\Psi\rangle = \sum_{i_1 \dots i_N = 0}^{1} c_{i_1 \dots i_N} |i_1 \dots i_N\rangle. \tag{5.18}$$

Such a state has 2^N possible complex coefficients $c_{i_1 \dots i_N}$. It is easy to understand that quantum systems with larger numbers of qubits are not simulatable on classical computers. A 52 qubit state can have up to $2^{52} \approx 4.5 \times 10^{15}$ non-zero complex coefficients. Beyond a certain point, it is not even possible to store the number of coefficients any more. Richard Feynman therefore suggested the use of quantum systems to simulate quantum systems.

5.5.3 Photonic two-input gates

For a universal set of quantum gates, a set of one-input gates plus one two-input gate, such as the CNOT gate, are required [77]. This universal set is the equivalent of the NOR or NAND gate in classical computing, which allows the implementation of any logical circuit just by combining NOR (or NAND) gates. While one-input gates can be implemented very efficiently for various qubit encoding schemes (such as path encoding, polarization encoding, etc.), photonic two-input gates are the main challenge for the implementation of photonic quantum computers. This is mainly caused by the bosonic character of photons, which prohibits the interaction that would be required for two-input gates. Interaction between photons occurs, for example, in non-linear materials, e.g. in the process of frequency doubling or in spontaneous parametric downconversion (SPDC) [79]. However, these non-linear effects are typically very inefficient, with efficiencies that only improve for very high light intensities. In a two-input gate in which two photons should interact, the light intensity is low by definition.

5.5.3.1 The CNOT gate

The controlled NOT (CNOT) gate is a two-input gate which can be used to build a universal set of gates for quantum computing in conjunction with a set of one-input gates. The symbol for the CNOT gate is shown in figure 5.28 and its truth table is shown in table 5.2. The gate's two inputs are called 'Control' and 'Target'. The 'Control' input is forwarded without any change to the 'Control' output. If the control input is $|0\rangle$, the 'Target' output equals the 'Target' input; if it is $|1\rangle$, the 'Target' output is the flipped version of the 'Target' input [78].

5.5.3.2 The probabilistic CNOT gate

A very interesting approach that only uses linear optical components to build probabilisitc gates was introduced by Knill *et al* in 2001 [2]; it is commonly known as the KLM scheme after the three authors, E Knill, R Laflamme, and G J Milburn. It uses the non-linearity introduced by the collapse of the wave function after a measurement is performed. Additionally, ancilla photons are introduced. Ancilla photons are 'helper photons' in the system that allow the determination of whether the gate operation was successful; they also increase the success rate of the gate. As the name 'probabilistic gate' already suggests, the successful operation of this gate is not guaranteed. However, using measurements of the ancilla photons, the success of the gate function can be determined.

Also in 2001, Pittman *et al* introduced a suggestion for an implementation of a probabilistic CNOT gate [80]. We will summarize and discuss the first part of their

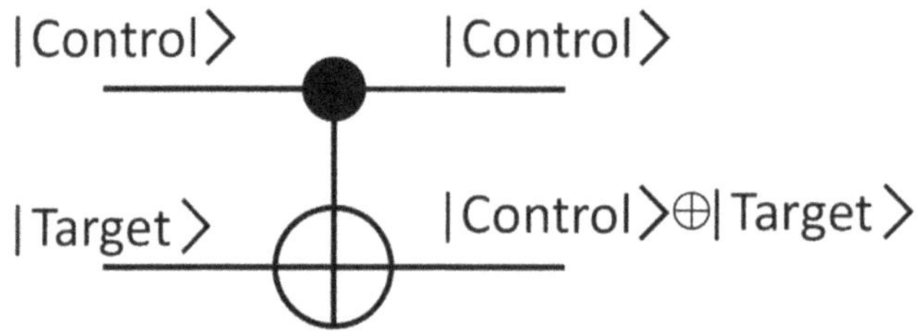

Figure 5.28. Symbol for a CNOT gate [78].

Table 5.2. Truth table of a CNOT gate [78]

	Input	Output	
Control	Target	Control	Target
$\|0\rangle$	$\|0\rangle$	$\|0\rangle$	$\|0\rangle$
$\|0\rangle$	$\|1\rangle$	$\|0\rangle$	$\|1\rangle$
$\|1\rangle$	$\|0\rangle$	$\|1\rangle$	$\|1\rangle$
$\|1\rangle$	$\|1\rangle$	$\|1\rangle$	$\|0\rangle$

implementation, namely a destructive probabilistic CNOT gate, as shown in figure 5.29, since this discussion helps in gaining a better understanding of the mechanisms of the KLM scheme. This CNOT gate is not deterministic, i.e. it will not be successful for every operation. However, the readings of the detectors 'H' and 'V' show whether the execution was successful. If the detector 'H' detects one and only one photon, then the gate worked as expected and the target output can be further used. If detector 'V' detects one and only one photon, the gate operation was unsuccessful, but it can be corrected by flipping the state of the target output. If zero or two photons are detected in the detectors, the gate operation was unsuccessful and needs to be discarded. This implementation has a success rate of 25% if the case with one and only one detection of 'V' is uncorrected, and a success rate of 50% if it is corrected. As well as only having a limited probability of successful operation, this implementation destroys the control qubit. After a discussion of this implementation, an extended version presented by the authors of [80] will be shown (but not discussed in detail), which does not destroy the control qubit.

This implementation works with polarization-encoded qubits. It uses polarizing splitters in two different bases, the H/V base and the diagonal base.

Qubits given in the H/V base can be derived in the diagonal base using the following simple equations, which were derived from figure 5.30 using simple trigonometry:

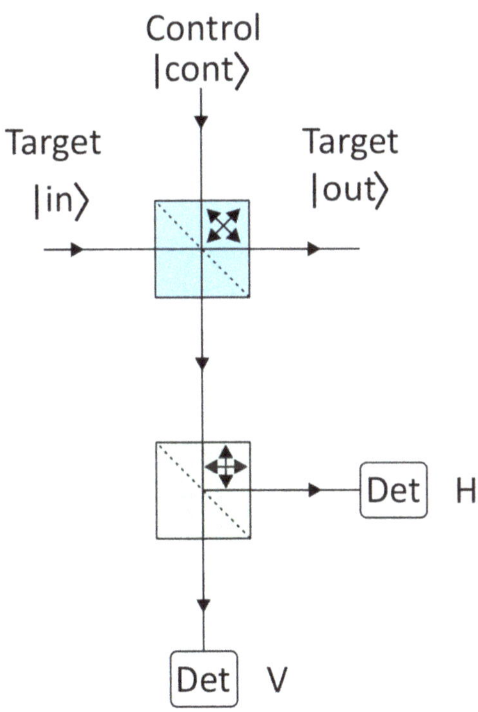

Figure 5.29. Probabilistic destructive CNOT gate [80].

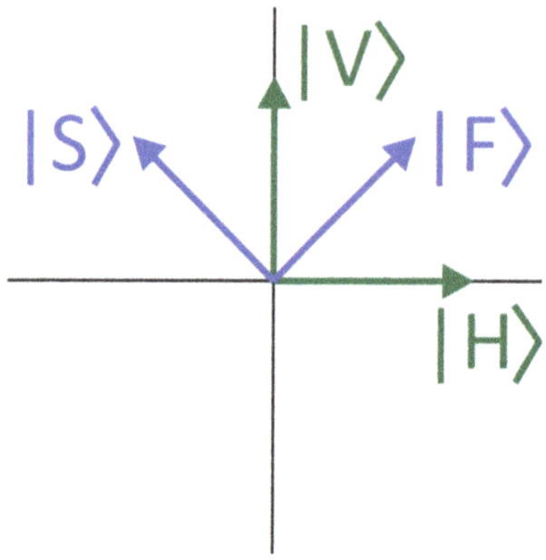

Figure 5.30. Orientations used for the two bases (the H/V base and the diagonal F/S base) [80].

$$|H\rangle = \frac{1}{\sqrt{2}}(|F\rangle - |S\rangle), \tag{5.19}$$

$$|V\rangle = \frac{1}{\sqrt{2}}(|F\rangle + |S\rangle), \tag{5.20}$$

where $|H\rangle$, $|V\rangle$, $|F\rangle$, and $|S\rangle$ correspond to horizontal, vertical, first diagonal, and second diagonal polarizations, respectively. The diagonal base (F/S base) can be derived from the H/V base as follows:

$$|F\rangle = \frac{1}{\sqrt{2}}(|H\rangle + |V\rangle), \tag{5.21}$$

$$|S\rangle = \frac{1}{\sqrt{2}}(-|H\rangle + |V\rangle). \tag{5.22}$$

Figure 5.31 shows the transmission/reflection characteristics of the polarizing beam splitters. The H and F polarizations are transmitted, while the V and S polarizations are reflected.

We will separate the derivation of the output signal into two cases, one in which the control input $|cont\rangle$ is $|0\rangle$ and one in which it is $|1\rangle$. We start with the case $|cont\rangle = |0\rangle = |H\rangle$ and assume the following general qubit at the input $|in\rangle$:

$$|in\rangle = \alpha|H\rangle + \beta|V\rangle \tag{5.23}$$

The first beam splitter is in the F/S base. Therefore, we need to write $|in\rangle$ and $|cont\rangle$ in the F/S base, using equations (5.19) and (5.20) in order to derive the states at the outputs of the splitter.

$$|in_i\rangle = \frac{\alpha}{\sqrt{2}}(|F_i\rangle - |S_i\rangle) + \frac{\beta}{\sqrt{2}}(|F_i\rangle + |S_i\rangle) \tag{5.24}$$

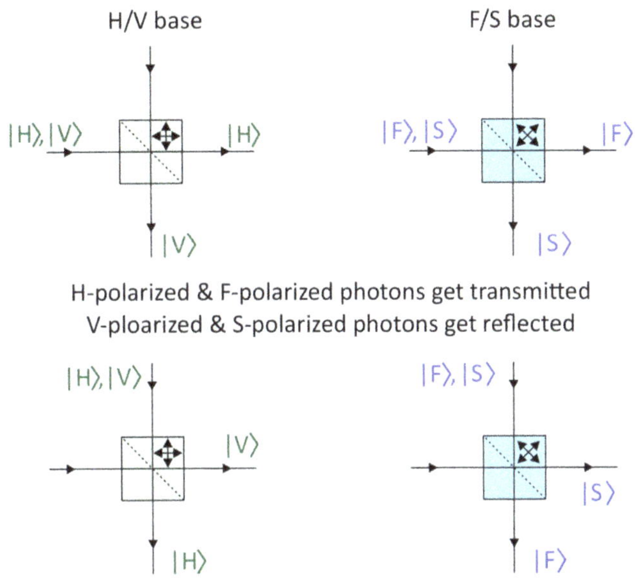

Figure 5.31. Characteristics of the splitters in the H/V base and the F/S base, as in [80].

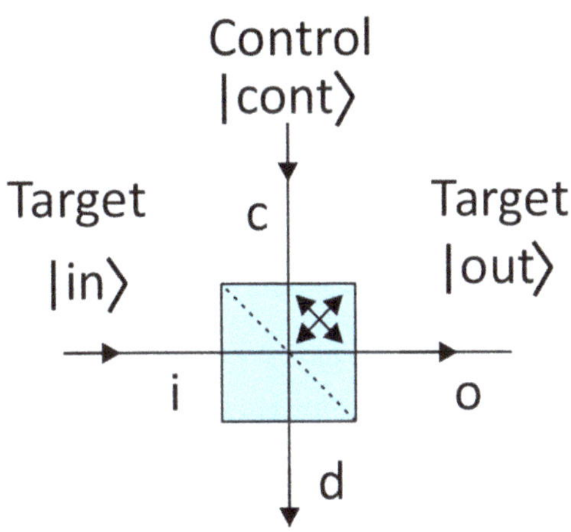

Figure 5.32. Modes i, c, o, and d at the first splitter [80].

$$|cont_c\rangle = \frac{1}{\sqrt{2}}(|F_c\rangle - |S_c\rangle) \tag{5.25}$$

The indices in the equation above indicate the spatial mode of the photon (i.e. which input or output it is in), as indicated in figure 5.32. The two-photon state $|\Psi_{\mathrm{i,\,c}}\rangle$ at the input is derived using the tensor product. We get:

$$|\Psi_{\mathrm{i,\,c}}\rangle = |in_{\mathrm{i}}\rangle \otimes |cont_{\mathrm{c}}\rangle \tag{5.26}$$

$$|\Psi_{i,c}\rangle = \frac{\alpha}{2}[|F_iF_c\rangle - |F_iS_c\rangle - |S_iF_c\rangle + |S_iS_c\rangle] + \frac{\beta}{2}[|F_iF_c\rangle - |F_iS_c\rangle + |S_iF_c\rangle - |S_iS_c\rangle] \tag{5.27}$$

To derive the two-qubit state at the output of the beam splitter, one needs to consider the transmission characteristics shown in figure 5.31. $|F\rangle$ photons are transmitted, meaning that from mode i they will end up in mode o, and from mode c they will end up in mode d. $|S\rangle$ photons are reflected, meaning that from mode i they will end up in mode d, and from mode c they will end up in mode o. Considering this, we obtain the following two-qubit state $\Psi_{o,\,d}$ at the output of the splitter:

$$|\Psi_{o,d}\rangle = \frac{\alpha}{2}[|F_oF_d\rangle - |F_oS_o\rangle - |S_dF_d\rangle + |S_dS_o\rangle] + \frac{\beta}{2}[|F_oF_d\rangle - |F_oS_o\rangle + |S_dF_d\rangle - |S_dS_o\rangle] \tag{5.28}$$

As already explained during the introduction to the probabilistic CNOT gate, only combinations in which one and only one photon is present at the output can have operated correctly. This is, for example, the case for the first term containing $|F_oF_d\rangle$, which corresponds to a state in which one F-polarized photon is in mode 'o', while another F-polarized photon is in mode 'd'. In contrast, for example, in the state containing $|F_oS_o\rangle$, there are two photons present in mode 'o,' meaning that we have two photons at the output, which is certainly an incorrect operation of the CNOT gate. For this given state, one of these two photons would be F-polarized, while the second is S-polarized.

We therefore can separate the terms in equation (5.28) into those in which exactly one photon is present at the output mode 'o' (i.e. the terms that only have one 'o' index) and those for which zero or two photons are present at the output mode 'o'. During a real experiment, one can detect whether this is the case by checking whether the detectors have detected zero or two photons[15]. In these cases, the gate operation was unsuccessful, and the result has to be discarded. These terms that have to be discarded are collected in $|\Psi_x\rangle$. Separating the terms leads to:

$$|\Psi_{o,d}\rangle = \frac{1}{2}[\alpha|F_oF_d\rangle + \alpha|S_dS_o\rangle + \beta|F_oF_d\rangle - \beta|S_dS_o\rangle] + |\Psi_x\rangle \tag{5.29}$$

$$|\Psi_x\rangle = \frac{1}{2}[-\alpha|F_oS_o\rangle - \alpha|S_dF_d\rangle - \beta|F_oS_o\rangle + \beta|S_dF_d\rangle] \tag{5.30}$$

Since the second splitter, the one that has the detectors at its outputs, is an H/V splitter, we need to write our two-qubit output state in the H/V base again. In a first step, we rewrite in the H/V base only the photons that are in the 'd' mode that leads to the detectors. In order to do so, we use equations (5.21) and (5.22):

[15] For this approach, photon-number-resolving detectors are important.

$$|\Psi_{\text{o, d}}\rangle = \frac{1}{2\sqrt{2}}[\alpha|F_\text{o}H_\text{d}\rangle + \alpha|F_\text{o}V_\text{d}\rangle - \alpha|H_\text{d}S_\text{o}\rangle + \alpha|V_\text{d}S_\text{o}\rangle \\ + \beta|F_\text{o}H_\text{d}\rangle + \beta|F_\text{o}V_\text{d}\rangle + \beta|H_\text{d}S_\text{o}\rangle - \beta|V_\text{d}S_\text{o}\rangle] + |\Psi_x\rangle \tag{5.31}$$

If we now separate the terms containing H_d (i.e. the states in which the H detector detects one photon) from those containing V_d (i.e. the states in which the V detector detects one photon) and if we then factor out $|H_\text{d}\rangle$ and $|V_\text{d}\rangle$, we obtain:

$$|\Psi_{\text{o, d}}\rangle = \frac{1}{2\sqrt{2}}[|H_\text{d}\rangle \otimes (\overbrace{\alpha|F_\text{o}\rangle - \alpha|S_\text{o}\rangle}^{=\alpha\sqrt{2}|H_\text{o}\rangle} + \overbrace{\beta|F_\text{o}\rangle + \beta|S_\text{o}\rangle}^{=\beta\sqrt{2}|V_\text{o}\rangle})] \\ + \frac{1}{2\sqrt{2}}[|V_\text{d}\rangle \otimes (\underbrace{\alpha|F_\text{o}\rangle + \alpha|S_\text{o}\rangle}_{=\alpha\sqrt{2}|V_\text{o}\rangle} + \underbrace{\beta|F_\text{o}\rangle - \beta|S_\text{o}\rangle}_{=\beta\sqrt{2}|H_\text{o}\rangle})] + |\Psi_x\rangle \tag{5.32}$$

In short, this results in:

$$|\Psi_{\text{o, d}}\rangle = \frac{1}{2}[|H_\text{d}\rangle \otimes (\alpha|H_\text{o}\rangle + \beta|V_\text{o}\rangle) + |V_\text{d}\rangle \otimes (\alpha|V_\text{o}\rangle + \beta|H_\text{o}\rangle)] + |\Psi_x\rangle \tag{5.33}$$

The first part of this equation shows the state at the output (i.e. $\alpha|H_\text{o}\rangle + \beta|V_\text{o}\rangle$) if one and only one photon is detected by detector 'H'. This state is the same as the input state, which is the correct output state if the control qubit is $|0\rangle$, as assumed in this first part. The second part of the equation shows the state at the output if one and only one photon is detected by the 'V' detector. In this case, the horizontal and vertical polarizations are flipped.

For the second case, in which the control qubit is $|1\rangle$, corresponding to $|V_\text{c}\rangle$, the derivation works in exactly the same way as for the first case. Therefore the step-by-step derivation will not be shown for this case, but only the final result. In this second case, the following two-qubit state is present at the output:

$$|\Psi_{\text{o, d}}\rangle = \frac{1}{2}[|H_\text{d}\rangle \otimes (\alpha|V_\text{o}\rangle + \beta|H_\text{o}\rangle) + |V_\text{d}\rangle \otimes (\alpha|H_\text{o}\rangle + \beta|V_\text{o}\rangle)] + |\Psi_x\rangle \tag{5.34}$$

The state at the output if detector 'H' detects one and only one photon is, in this case, flipped (it is $\alpha|V_\text{o}\rangle + \beta|H_\text{o}\rangle$) compared to the input state shown in equation (5.23). This is exactly the state we need at the output, since, for a $|1\rangle$ at the control input, the input state should be flipped. If the output of detector 'V' is one and only one photon, the state at the output is $\alpha|H_\text{o}\rangle + \beta|V_\text{o}\rangle$, which needs to be flipped to correct it.

We can finally summarize [80]:

- If zero or two photons are detected by the detectors, the gate operation failed and the result needs to be discarded. The probability that this will happen is 50%.
- If one and only one photon is detected by detector 'H', then the output of this gate is correct and can be used. This is true for $|0\rangle$ as well as $|1\rangle$ at the control input and therefore also for all possible superpositions. The probability that this is the case is 25%.

- If one and only one photon is detected by detector 'V', then the output of this gate is flipped and needs to be corrected. If forward error correction is used (i.e. the output qubit is flipped if detector 'V' detects one photon), this case can also be used. This is true for $|0\rangle$ as well as $|1\rangle$ at the control input and therefore also for all possible superpositions. The probability that this is the case is 25%.
- If no forward error correction is used, this gate has a success rate of 25%; if forward error correction is used, the success rate increases to 50%.

Pittman *et al* [80] also presented an extended version of the gate, as shown in figure 5.33. In the lower part of the structure, the destructive CNOT that was discussed above is visible. This extended version also includes helper photons (ancilla photons), which are provided by an entangled photon-pair source. This source generates photon pairs that are polarization entangled in such a way that if a photon in H (V) polarization is detected in mode 'a,' the photon in mode 'b' has the same H (V) polarization. In this extended version, the control qubit is not destroyed.

This gate works correctly if one and only one photon is detected by detector 'F'. In this case, the correct control qubit is present at the output as well as at node b,

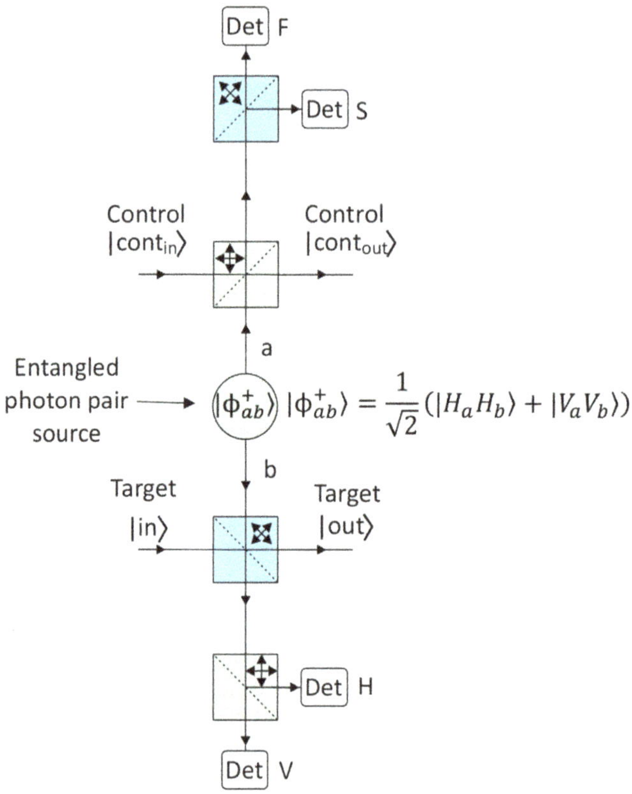

Figure 5.33. Probabilistic non-destructive CNOT gate according to [80].

where it acts as a control qubit for the destructive CNOT gate below. Additionally, one and only one photon needs to be detected by detector 'H'. In this case, the complete CNOT gate operation is successful. The success rate for this gate is 6.25%. If forward error correction is used, then the cases in which one and only one photon is detected by the detectors 'S' and 'V' can also be used, and the success rate increases to 25%.

An experimental implementation of such a gate was published by Gasparoni *et al* in 2004 [81], proving its feasibility. In 2018, Zeuner *et al* presented a CNOT gate in which parts of the gate were integrated onto a photonic chip, paving the way for future scalability demands.

It can be shown that increasing numbers of ancilla photons cause the success rate of probabilistic gates to theoretically reach values arbitrarily close to 100% [2, 82]. However, the use of a large number of photons increases the challenges regarding photon sources and detection. If, for example, N photons need to be detected at the same time in N detectors, the probability P_{NP} that all N photons are detected is:

$$P_{NP} = PDE^{N} \tag{5.35}$$

if only the photon detection efficiency (PDE) is considered. The peak PDE of state-of-the-art counting modules that incorparate SPADs reaches about 70% [17]. For the non-destructive probabilistic CNOT gate shown in figure 5.33, two photons need to be detected at the same time (one by the H/V detectors and one by the F/S detectors). The probability that both photons are really detected if a single detector has a PDE of 70% is only 49%. If the number of ancilla photons is increased in order to improve the success rate of the quantum gate, a PDE of close to 100% is crucial in order to avoid increasing the error rate at the detector side considerably.

5.5.4 Cluster states

Another promising method for photonic quantum computing is the utilization of cluster states. These states are highly entangled states, which ideally contain many photons. Preparing these states by bringing them into a known initial state and measuring single photons of these cluster states in specific bases, allows the remaining photons to be brought into the state they would be in if they were passing, for example, two-input quantum gates. The functionality of the 'quantum circuit' is defined by the initial cluster state and the specific sequence of measurements [83].

The advantage of this approach compared to the KLM scheme is that the success probability can actually reach 100%. In addition, it typically requires considerably fewer measurements than the KLM scheme. Its drawback is that it is a one-way quantum computing scheme. After a photon of the initial cluster has been measured, it is no longer available for further steps. Therefore, for complex quantum circuits, clusters with large numbers of photons and a high degree of entanglement are required. Nevertheless, in [83], Walther *et al* demonstrated a universal set of quantum gates using cluster states.

5.5.5 Requirements for single-photon detectors in quantum computing

Currently, the most critical parameter of single-photon detectors in photonic quantum computing is their PDE. For a scalable system, a PDE close to 100% is crucial, which is not currently feasible for SPADs. SNSPDs can approach this value much more closely, but are bulkier and more expensive due to their operation at cryogenic temperatures. Furthermore, they cannot be integrated as efficiently as SPADs. SPADs can be integrated with the quenching circuitry as well as with additional intelligence. It is much more difficult to integrate the readout electronics in SNSPDs due to their operation near 2 K. Consequently, it is also more difficult to make arrays with large numbers of pixels. While SPAD arrays have rapidly approached the megapixel array size [42], SNSPDs cannot currently be manufactured at such high pixel numbers.

Figure 5.34 visualizes the importance of a high PDE; it shows the detection probabilities of four different PDEs according to equation (5.35) for varying numbers of photons that have to be detected. Please bear in mind that the maximum value of $N = 100$ shown here is still very low and does not allow powerful universal quantum computing, since the number of qubits is considerably lower than the number of detectors, especially if the linear approach of [2] is used. Nevertheless, this figure shows that even for a PDE of 70% (which is the PDE of commercially available counting modules using silicon SPADs such as the module from Laser Components [15]), the probability of detecting all N photons drops to less than 10^{-10}% if 'only' 78 photons have to be detected simultaneously. For SPADs in the telecom wavelength regime at around 1550 nm, the situation is even worse. As they currently achieve a PDE of no more than 36% [25], they are not yet suitable for large photonic quantum systems. Although they are quite bulky, SNSPDs seem to be the better choice for large-scale photonic quantum computing experiments at the moment. The SNSPD described in [24] achieved a system detection efficiency of

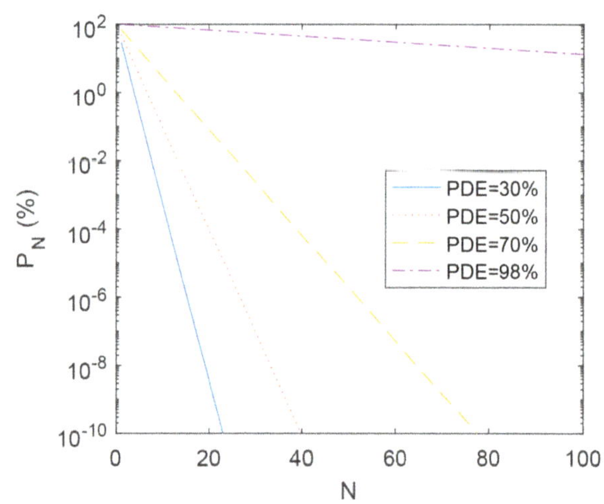

Figure 5.34. Probability of detecting N single photons in N detectors depending on N, for four different PDEs.

98%, which yields a detection probability that is still greater than 10% for N=100 photons.

5.6 Ghost imaging

While the first ghost imaging experiments were performed using photon pairs from SPDC, the basic principle of ghost imaging also works without quantum effects [6] and just relies on spatial correlation between two light fields. However, utilizing the quantum nature of photons can result in advantages, compared to the classical version. This advantage can, for example, be an improved signal-to-noise ratio (SNR) in the recorded image [84], an improved spatial resolution in the near field [85], or an improved field of view in the far field [85].

In this section, we will discuss the basic principle of ghost imaging, a selection of its potential applications, and the requirements for single-photon detectors used for ghost imaging.

SPDC is also used in photonic quantum computing and quantum simulation for the generation of heralded photons. One photon of the pair is used for the experiment and is typically called the ‘signal’ photon, while the second photon is utilized to show or trigger the presence of the ‘signal’ photon in the experiment. This second photon is typically called the ‘idler’ photon. In SPDC, the spatial correlation results from the conservation of momentum, i.e. the sum of the momentums of the outgoing photons equals the momentum of the incoming photon [79]. Additionally, the conservation of energy results in the following equation for the frequencies of the photons:

$$\omega_{\mathrm{p}} = \omega_{\mathrm{s}} + \omega_{\mathrm{i}}, \tag{5.36}$$

where ω_{p} is the frequency of the pump photon, and ω_{s} and ω_{i} are the frequencies of the signal photon and the idler photon, respectively [79]. In ghost imaging, two different types of detector are typically used to detect two spatially correlated light fields, as depicted in figure 5.35. In the first type, a bucket detector is located in the imaging path, where the object to be imaged is also placed. In the second type, a pixel detector is placed in an optical path that does not interact at all with the object to be imaged. The basic idea behind this imaging technique is to utilize the abovementioned spatial correlation between the two light fields. Without any information from the bucket detector, the pixel array detector would just image the SPDC source. However, if only those detections in the pixel array are considered, when there is a simultaneous detection in the bucket detector, the image can be reconstructed. For a better understanding, let us take a close look at figure 5.35 again. In this example, only the white parts of the object transmit photons to the bucket detector. If the photon hits a black part, it does not reach the bucket detector. Since the two light fields are spatially correlated, the position of a photon detected by the pixel array depends on the path that the second photon took that interacts with the object. Therefore, if only those photons which have a simultaneous detection in the bucket detector are considered in the pixel array, the shape of the transparent part of the object can be reconstructed.

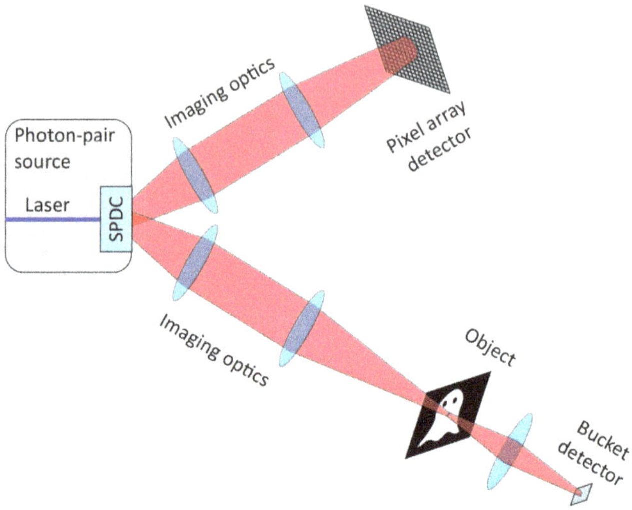

Figure 5.35. Basic principle of ghost imaging, after [6]. Two spatially correlated light fields, for example, those generated by an SPDC photon-pair source are used. The one that passes the object to be imaged is only detected by a single-pixel detector (bucket detector), while the second light field, which does not directly interact with the object, is detected by a pixel array. The image can be reconstructed using coincidence detection.

Using SPDC for ghost imaging has some great advantages. Since the two photons are generated at the same instant, the photon in the bucket detector is utilized to gate the pixel array. This can have a large impact on the noise in the recorded image. The portion of noise introduced, for example, by the dark count rate of the pixel array detector, can be considerably reduced by coincidence measurement between the bucket and the array detector, which is necessary for ghost imaging anyway. Please note that this statement would still hold if the object is placed in the path of the light field propagating toward the pixel array detector. The bucket detector could still be used for coincidence detection to gate the pixel array. In [84], Morris *et al* investigated such a setup (called heralded imaging) as well as a ghost-imaging setup and compared it with a classical imaging approach without coincidence detection. They showed that both heralded imaging and ghost imaging require considerably fewer photons per pixel in order to reconstruct a reasonable picture of the object, compared to classical imaging. Using heralded imaging or ghost imaging, they were able to reconstruct reasonable images with less than one photon per pixel on average.

The SPDC crystal can be designed in such a way that the frequencies, and therefore also the wavelengths of the signal and the idler photon, are far from each other [86]. The two frequencies just need to follow equation (5.36). This can be achieved by using a completely different wavelength range for the bucket detector and the array detector. If one is interested in imaging in a wavelength range in which pixel arrays are unavailable or very expensive, ghost imaging can offer a solution. The SPDC crystal needs to be prepared in such a way that the one photon, which will be sent through the object, is in the desired wavelength range for the imaging application, while the second photon is in a wavelength range that can be effectively detected by available and affordable pixel arrays. Aspden *et al* used this approach to probe an object using 1550 nm light, while the photons of the light field propagating toward the camera had a wavelength of 460 nm [86]. This approach can also be of interest for spectroscopic applications.

Many additional variants of ghost imaging approach exist, as presented in [87], for example; some of them only require one single-pixel detector. These approaches are called computational ghost imaging. One option is to replace the pixel array detector by, for example, a digital micromirror device (DMD) or a spatial light modulator (SLM) to spatially modulate the light field that passes the object before it hits the single-pixel detector [88, 89].

5.6.1 Requirements for single-photon detectors in ghost imaging

Classical ghost imaging relies on a coincidence measurement between a pixel array detector and a single-pixel detector (also called a bucket detector). For the coincidence measurement, a high timing resolution and therefore a low timing jitter are beneficial. This high-resolution timing allows the object to be imaged using a high photon-pair rate. Additionally, high-resolution timing allows the removal of the portion of dark counts for which no coincident detection was found in the second detector. This helps to reduce the noise in the image.

For the pixel array detector, a high pixel count is beneficial. Since megapixel SPAD arrays were only presented recently [42], many ghost imaging experiments utilized intensified charge coupled device (ICCD) cameras instead. These ICCD cameras contain a microchannel plate and a phosphor screen that preamplify the signal before it hits the CCD camera chip. This preamplifier can be utilized as a fast shutter and can therefore be used for the coincidence detection required in ghost imaging [90]. However, a larger degree of integration will potentially lead to considerably cheaper and smaller megapixel SPAD arrays compared to ICCD cameras in the future.

A decent PDE is helpful, since two photons have to be detected simultaneously. However, it is not as critical as for quantum computing or boson sampling, in which many more photons have to be detected at the same time.

One of the advantages of ghost imaging is that the photons that interact with the object can have a wavelength that is not detectable by the pixel array detector. Only the bucket detector needs to be capable of detecting the photons that are interacting with the object. This allows, for example, the use of a silicon SPAD array in a CMOS process for the pixel array, with all the advantages that integration in CMOS has to offer, while a specialized single-photon detector can be used for the bucket detector. This specialized detector could be, for example, an InGaAs SPAD capable of detecting 1550 nm. In some applications, it could even make sense to use a SNSPD for the bucket detector, as it offers high sensitivity, a low dark count rate, and typically high timing resolution combined with a broad range of operational wavelengths. Combining this SNSPD bucket detector with a silicon SPAD array offers the best of both worlds.

5.7 Super-resolution microscopy

In recent years, great progress has been achieved in super-resolution microscopy, which allows optical microscopy below the Abbe limit, which is defined as:

$$d = \frac{\lambda}{2n\sin(\alpha)} = \frac{\lambda}{2NA}, \tag{5.37}$$

where d corresponds to the resolution, λ to the wavelength of the used light, n to the refractive index of the material the object and the lens are immersed in, α corresponds to half of the opening angle of the lens, and NA corresponds to the numerical apperture [91]. With modern lenses and by immersing the object in oil, a numerical aperture of ~1.4 is feasible, corresponding to resolutions of approximately $\lambda/2.8$. In practice, very often, the Abbe limit is estimated as half of the used wavelength.

In order to improve the resolution, e.g. shorter wavelengths can be used. However, optics for these wavelengths get more and more complicated the shorter the wavelengths get. Furthermore, the use of photons with shorter wavelengths and therefore higher photon energy might damage the sample.

Therefore, other approaches have been investigated in order to allow microscopy of smaller structures to take place while still using visible light. One very successful approach is single-molecule localization microscopy, where e.g. the locations of small light emitters (fluorophores) that are attached to the sample are estimated by Gaussian fitting. This approach will be discussed in the next part. After that, another approach, namely quantum imaging, will be discussed. In quantum imaging, the correlation between entangled photons is utilized in order to improve the achievable resolution. It can be shown that for N entangled photons, the resolution can be improved by the factor of N, since the de Broglie wavelength of an ensemble of N entangled photons is reduced by a factor of N compared to the wavelength of a single photon.

While single-molecule localization microscopy (SMLM) approaches, such as stochastic optical reconstruction microscopy (STORM), have already found their way into applications, quantum imaging is still at a very early stage. However, very promising proof-of-concept experiments for this approach also exist [8].

5.7.1 Single-molecule localization microscopy

Most implementations of single-molecule localization microscopy (SMLM) are not real quantum applications, since they do not utilize quantum mechanics in order to go beyond classical physics. However, they use very smart approaches which allow imaging below the classical Abbe limit and which require the detection of very few photons (in the range of a few hundred) [92]. Additionally, we will compare this method with super-resolution quantum microscopy, which really utilizes quantum effects.

In SMLM, structures are imaged utilizing point light emitters attached to the structure to be imaged. In the process of imaging, only a few of these point emitters emit light at a time. This light emission can be captured by a microscope as a blurry spot, or, more exactly, as a Gaussian distribution of detected photons. The position of the light emitter can be estimated by fitting a Gaussian distribution to the measured blurry spot. This allows an estimation of the light emitter's position that is far more precise than the wavelength of the light used. The positional resolutions are typically in the range of 20–50 nm [7].

Let us take a closer look at stochastic optical reconstruction microscopy (STORM), a very successful SMLM approach. In STORM, the structures to be imaged need to be prepared by attaching photo-switchable molecules to them, which can be optically switched to a fluorescent state. In [93], a very detailed description of this method is presented, which will be summarized here. The photo-switchable molecules can be comprised, for example, of pairs of fluorophores; one is the 'activator,' the second is the 'reporter.' If the activator is excited by a laser pulse with a specific wavelength, the nearby reporter is switched from its dark state to its fluorescent state, in which it can emit several hundred to several thousand photons when excited by a laser with a wavelength that is typically longer than that used for the activator.

The image is recorded in many steps, as depicted in figure 5.36. After the sample has been prepared by the attachment of the photo-switchable molecules, the measurement can begin. A small fraction of the activators is excited by a weak laser pulse. The excitation of the activators is typically a purely random process. When an activator is excited, the corresponding neighbouring reporter becomes fluorescent. A second laser excites the switched-on reporters, thereby resulting in fluorescent emission from these molecules. When activated, these reporter molecules can emit several hundred to a few thousand photons when excited by a laser before falling back into the dark state again. These photons are captured by a classical microscope. The recorded image is composed of clouds of recorded photons with a Gaussian distribution, as depicted in the central column of figure 5.36. The number of activators that are excited needs to be sufficiently small that the probability is high that most of the Gaussian photon distributions are non-overlapping. If this is the case, the location of the emitter can be estimated by a Gaussian fit to this distribution; the most probable location of the emitter is at the peak of the Gaussian shape. These most probable locations of the currently recorded Gaussian distributions are indicated by yellow crosses in the right column of figure 5.36. The positions of these emitters are then stored, as indicated by blue crosses in the same figure. The procedure is then repeated. Another set of activators is excited, which results in another set of reporters being switched on. Their positions are also localized. This procedure is repeated several hundred to several hundred thousand times, until the positions of a sufficiently large number of emitters are localized, thereby reconstructing the object to be imaged. The bottom part of figure 5.36 depicts such an image.

If the number of activated reporters per frame is increased, the number of frames required for image reconstruction can be reduced. In [94], quantum optic methods were exploited to extract the locations of densely packed emitters by analyzing

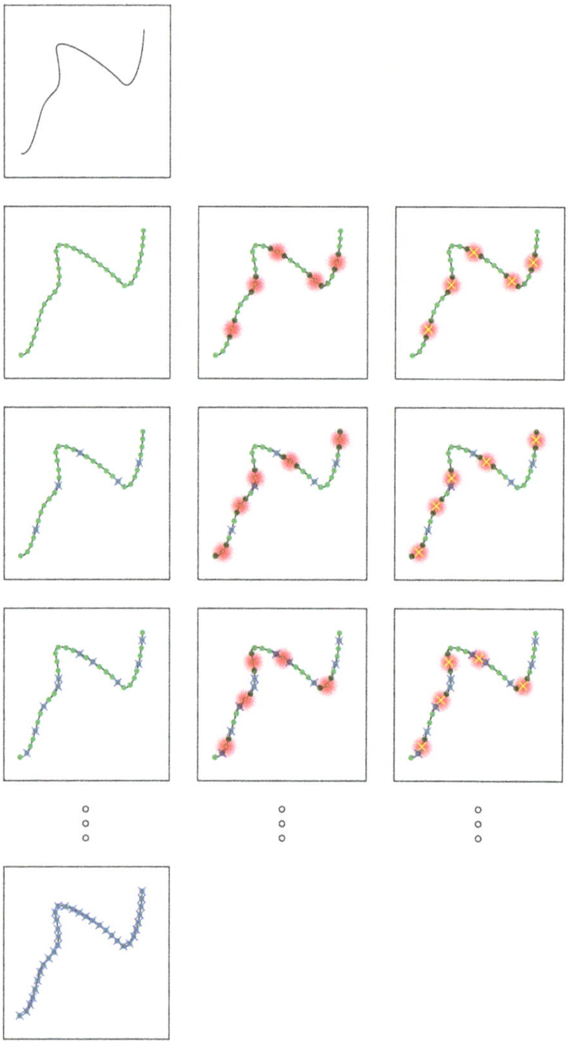

Figure 5.36. Basic principle of stochastic optical reconstruction microscopy (STORM), a single-molecule localization microscopy (SMLM) method [93]. The structure under investigation is depicted in the top part. Photo-switchable fluorophores are attached to the structure in the image below. Some of these fluorophores are activated by a weak laser pulse and then start fluorescent emission if illuminated by a second laser. The centers of the light spots of the fluorescent spots are estimated. The steps from the activatation of some fluorophores to the estimation of the emission centers are repeated until the whole structure is revealed.

spatially resolved time streams of detected photons. M Aßmann showed that for a density of up to 125 emitters per μm, a resolution (i.e. a localication accuracy) of less than 30 nm can be achieved. However, while classical STORM can work, for example, with electron multiplying charge-coupled device (EMCCD) cameras [93], this quantum optically enhanced method relies on the availability of fast single-photon detectors, such as SPAD arrays, since it requires excellent timing resolution.

In contrast to ICCD cameras, EMCCD cameras do not utilize a preamplifier composed of a microchannel plate and a phosphor screen, but instead use electrical amplification in the readout phase utilizing impact ionization [95].

5.7.2 Super-resolution quantum microscopy

The de Broglie wavelength of N entangled photons with a wavelength of λ is given by λ/N. This was experimentally proven in [96] for an entangled photon pair and in [97] for a four-photon state.

This reduction of the wavelength of the photon ensemble can be utilized for super-resolution quantum microscopy. Due to the reduced de Broglie wavelength, the resolution of a microscope can be improved by a factor of N if N-fold entangled photons are used. This improved resolution, i.e. the classical limit improved by the factor $1/N$, is called the Heisenberg limit. A big advantage of super-resolution quantum microscopy compared to SMLM is that the preparation of samples is much easier, since no fluorescent emitters need to be attached to the sample.

The European project SUPERTWIN had the goal of developing a proof of concept for such super-resolution quantum microscopy [98]. There are two main technological challenges.

First, for a large improvement of the achievable resolution, the entangled state needs to contain a large number of photons (i.e. N needs to be large). In [99], Zhong *et al* presented a photon source for entangled 12-photon states. While the authors claimed to have built a very efficient source, the rate of these 12-photon states was only approximately one per hour. This rate is obviously insufficient for an efficient imaging approach—the recording of the image would take too long. The achievable rates increase considerably for entangled states containing fewer photons, but reducing N also reduces the improvement in resolution.

The second technological challenge is that all the photons of the entangled state need to be detected. Coincidence detection can be utilized in order to discover which detected photons belong to which entangled state. Photons that are detected simultaneously have a high probability of originating from the same N-photon state. The image is reconstructed from these detections by an optical centroid measurement, introduced by Tsang *et al* in [100]. Since the high resolution should be paired with a reasonable field of view to be applicable in praxis, arrays are required that have a large number of pixels capable of performing coincident detection.

In [98], a proof of concept was presented using a 8×16 pixel SPAD array that allowed coincidence detection, and a path toward a 100 kilopixel sensor was discussed, as part of the SUPERTWIN roadmap.

Toninelli *et al* demonstrated an improvement of the achievable resolution below the Abbe limit for a two-photon state [101]. However, their resolution improvement was still quite far from the Heisenberg limit. Instead of an SPAD array, they used an EMCCD, which resulted in a long measurement time per image.

In 2018, Unternährer *et al* presented a proof of concept experiment that achieved a resolution at the Heisenberg limit [8], which was one of the important outcomes of SUPERTWIN. In their experiment, N was still limited to a value of two. As a photon source, they used a non-linear crystal pumped at 405 nm to generate SPDC photon pairs. The detector was a 32 × 32 SPAD array that reached a frame rate of 800 kHz. Coincidence detection was utilized to reduce the noise caused by dark counts in the recorded image.

Due to the difficulty of generating entangled states with larger numbers of photons and the requirements on the detector side, super-resolution quantum microscopy is still not as mature as SLML, and it is still at an experimental level.

5.7.3 Requirements for single-photon detectors in super-resolution microscopy

For single-molecule localization microscopy, there are two main critical factors. For practical applications, pixel arrays with a sufficient number of pixels are crucial. A high PDE is important, since the spatial resolution is directly related to the number of photons received per emitter. The higher the number of received photons, the better the quality of the Gaussian fit on average and the better the localization of the emitter.

For super-resolution quantum imaging, the PDE is even more critical. Since the resolution improvement compared to classical microscopy relies on the simultaneous detection of N entangled photons (where N should be as large as possible), a high PDE is beneficial for the single-photon detectors required here.

Since a pixel array is required, SPADs have a clear advantage compared to SNSPDs. However, the requirement for a pixel array introduces challenges in achieving a high PDE. To reach large pixel numbers, the pixel size has to be small. A small pixel size leads to a reduced fill factor—first, to prevent crosstalk between neighbouring SPADs, and second, because each SPAD needs to be quenched. Additionally, the chip needs some kind of intelligence, ideally integrated into each pixel, in order to perform coincidence detection of the N entangled photons. Currently, two successful approaches are utilized in order to achieve a high fill factor at a small pixel pitch: first, the use of microlenses that concentrate the incoming light on the active area of the pixel [38, 39]; second, 3D integration is used, in which the SPAD is integrated into one chip (which even can be process optimized to improve the key parameters of the SPAD), and a second chip contains the quenching and timing circuitry [40, 41].

If the number of pixels is large, detector noise, such as dark counts and afterpulsing become critical. Their influence can be reduced to a certain level by synchronizing the detection with the entangled photon source. However, even this will not remove all of the noise. Nevertheless, the dark count rate of the SPADs can be reduced by orders of magnitude by cooling the chip [74].

Since the measurement principle relies on coincidence detection, the timing jitter is also critical. The timing jitter and the dead time of the SPAD directly determine the maximum possible rate at which the photon source can be operated, and therefore also determine the time required to take one microscopic image. This will become especially important when efficient and bright entangled photon sources with large numbers of entangled photons become available. Currently, the efficiency of the source determines the measurement time much more than the performance of the detector.

References

[1] Zhong H-S *et al* 2020 Quantum computational advantage using photons *Science* **370** 1460–3

[2] Knill E, Laflamme R and Milburn G J 2001 A scheme for efficient quantum computation with linear optics *Nature* **409** 46–52

[3] Lucamarini M, Yuan Z L, Dynes J F and Shields A J 2018 Overcoming the rate-distance limit of quantum key distribution without quantum repeaters *Nature* **557** 400–3

[4] Chen J-P *et al* 2020 Sending-or-not-sending with independent lasers: Secure twin-field quantum key distribution over 509 km *Phys. Rev. Lett.* **124** 070501

[5] Zhang Q, Xu F, Chen Y-A, Peng C-Z and Pan J-W 2018 Large scale quantum key distribution: challenges and solutions *Opt. Express* **26** 24260–73

[6] Padgett M J and Boyd R W 2017 An introduction to ghost imaging: quantum and classical *Philos. Trans. R. Soc.* A **375** 20160233

[7] Lelek M, Gyparaki M T, Beliu G, Schueder F, Griffié J, Manley S, Jungmann R, Sauer M, Lakadamyali M and Zimmer C 2021 Single-molecule localization microscopy *Nat. Rev. Methods Primers* **1** 39

[8] Unternährer M, Bessire B, Gasparini L, Perenzoni M and Stefanov A 2018 Super-resolution quantum imaging at the heisenberg limit *Optica* **5** 1150–4

[9] Natarajan C M, Tanner M G and Hadfeld R H 2012 Superconducting nanowire single-photon detectors: physics and applications *Supercond. Sci. Technol.* **25**

[10] Gol'tsman G N, Okunev O, Chulkova G, Lipatov A, Semenov A, Smirnov K, Voronov B, Dzardanov A, Williams C and Sobolewski R 2001 Picosecond superconducting single-photon optical detector *Appl. Phys. Lett.* **79** 705–7

[11] Semenov A D, Gol'tsman G N and Korneev A A 2001 Quantum detection by current carrying superconducting film *Physica C: Superconductivity* **351** 349–56

[12] Yang J K W, Kerman A J, Dauler E A, Anant V, Rosfjord K M and Berggren K K 2007 Modeling the electrical and thermal response of superconducting nanowire single-photon detectors *IEEE Trans. Appl. Superconduct.* **17** 581–5

[13] Hadfield R H 2009 Single-photon detectors for optical quantum information applications *Nature Photon.* **3** 696–705

[14] Haitz R H 1965 Mechanisms contributing to the noise pulse rate of avalanche diodes *J. Appl. Phys.* **36** 3123–31

[15] https://www.lasercomponents.com/fileadmin/user_upload/home/Datasheets/lc-photon-counter/count-series.pdf. [Last accessed on 2021-08-23].

[16] http://www.hamamatsu.com/resources/pdf/ssd/c11202series_kacc1207e.pdf. [Last accessed on 2021-08-23].

[17] http://www.excelitas.com/product-category/single-photon-counting-modules. [Last accessed on 2021-08-23].

[18] Dervić A, Steindl B, Hofbauer M and Zimmermann H 2019 High-voltage active quenching and resetting circuit for SPADs in 0.35 μm CMOS for raising the photon detection probability *Optical Eng.* **58** 1–4

[19] Dervić A, Hofbauer M, Goll B and Zimmermann H 2021 High slew-rate quadruple-voltage mixed-quenching active-resetting circuit for SPADs in 0.35-μm CMOS for increasing PDP *IEEE Solid-State Circ. Lett.* **4** 18–21

[20] https://www.corning.com/media/worldwide/coc/documents/Fiber/SMF-28%20Ultra.pdf [Last accessed on 2021-08-30]

[21] https://www.thorlabs.de/_sd.cfm?fileName=19712-S01.pdf&partNumber=S630-HP [Last accessed on 2021-08-30].
[22] Zimmermann H 2010 *Integrated silicon optoelectronics* (Berlin: Springer) https://doi.org/10.1007/978-3-642-01521-2
[23] https://marketing.idquantique.com/acton/attachment/11868/f-023b/1/-/-/-/-/ID281_Brochure.pdf [Last accessed on 2021-09-07]
[24] Reddy D V, Nerem R R, Sae Woo Nam, Mirin R P and Varun 2020 Superconducting nanowire single-photon detectors with 98% system detection efficiency at 1550nm *Optica* **7** 1649–53
[25] Zhang J *et al* 2021 Triple-mesa InGaAs/InAlAs single-photon avalanche diode array for 1550nm photon detection *2021 Conf. on Lasers and Electro-Optics (CLEO)* (IEEE) 1–2
[26] Wollman E E *et al* Oct 2017 UV superconducting nanowire single-photon detectors with high efficiency, low noise, and 4 k operating temperature *Opt. Exp.* **25** 26792–801
[27] Zhang W J, Yang X Y, Li H, You L X, Lv C L, Zhang L, Zhang C J, Liu X Y, Wang Z and Xie X M feb 2018 Fiber-coupled superconducting nanowire single-photon detectors integrated with a bandpass filter on the fiber end-face. *Supercond. Sci. Technol.* **31** 035012
[28] https://marketing.idquantique.com/acton/attachment/11868/f-0234/1/-/-/-/-/ID230_Brochure.pdf. [Last accessed on 2021-09-07].
[29] Amri E, Boso G, Korzh B and Zbinden H Dec 2016 Temporal jitter in free-running InGaAs/InP single-photon avalanche detectors *Opt. Lett.* **41** 5728–31
[30] Burenkov V, Xu H, Qi B, Hadfield R H and Lo H-K 2013 Investigations of afterpulsing and detection efficiency recovery in superconducting nanowire single-photon detectors *J. Appl. Phys.* **113** 213102
[31] Korzh B *et al* 2020 Demonstration of sub-3 ps temporal resolution with a superconducting nanowire single-photon detector *Nature Photon.* **14** 250–5
[32] Nolet F, Parent S, Roy N, Mercier M-O, Charlebois S A, Fontaine R and Pratte J-F 2018 Quenching circuit and SPAD integrated in CMOS 65 nm with 7.8 ps fwhm single photon timing resolution *Instruments* **2** 19
[33] Tosi A, Calandri N, Sanzaro M and Acerbi F 2014 Low-noise, low-jitter, high detection efficiency InGaAs/InP single-photon avalanche diode *IEEE J. Sel. Topics Quantum Electron.* **20** 192–7
[34] Zhang W, Huang J, Zhang C, You L, Lv C, Zhang L, Li H, Wang Z and Xie X 2019 A 16-pixel interleaved superconducting nanowire single-photon detector array with a maximum count rate exceeding 1.5 ghz *IEEE Trans. Appl. Supercond.* **29** 1–4
[35] Vetter A *et al* 2016 Cavity-enhanced and ultrafast superconducting single-photon detectors *Nano Lett.* **16** 7085–92
[36] Steindl B, Hofbauer M, Schneider-Hornstein K, Brandl P and Zimmermann H 2018 Single-photon avalanche photodiode based fiber optic receiver for up to 200 Mb/s *IEEE J. Sel. Topics Quantum Electron.* **24** 1–8
[37] Dixon A R, Dynes J F, Yuan Z L, Sharpe A W, Bennett A J and Shields A J 2009 Ultrashort dead time of photon-counting InGaAs avalanche photodiodes *Appl. Phys. Lett.* **94** 231113
[38] Intermite G *et al* 2015 Fill-factor improvement of Si CMOS single-photon avalanche diode detector arrays by integration of diffractive microlens arrays *Opt. Express* **23** 33777–91
[39] Pellegrini S, Rae B, Pingault A, Golanski D, Jouan S, Lapeyre C and Mamdy B 2017 Industrialised SPAD in 40 nm technology *2017 IEEE Int. Electron Devices Meet. (IEDM)* 16.5.1–6.5.4

[40] Charbon E, Bruschini C and Lee M-J 2018 3D-stacked CMOS SPAD image sensors: Technology and applications *2018 25th IEEE Int. Conf. on Electronics, Circuits and Systems (ICECS)* pp 1–4

[41] Henderson R K, Johnston N, Hutchings S W, Gyongy I, Tarek Al Abbas, Dutton N, Tyler M, Chan S and Leach J 2019 5.7 A 256 × 256 40 nm/90 nm CMOS 3D-stacked 120 dB dynamic-range reconfigurable time-resolved SPAD imager *2019 IEEE International Solid-State Circuits Conference - (ISSCC)* pp 106–8

[42] Morimoto K, Ardelean A, Wu M-L, Ulku A C, Antolovic I M, Bruschini C and Charbon E 2020 Megapixel time-gated SPAD image sensor for 2D and 3D imaging applications *Optica* **7** 346–54

[43] Wollman E E, Verma V B, Lita A E, Farr W H, Shaw M D, Mirin R P and Nam S W 2019 Kilopixel array of superconducting nanowire single-photon detectors *Opt. Express* **27** 35279–89

[44] Wang H, Li H, You L, Wang Y, Zhang L, Yang X, Zhang W, Wang Z and Xie X 2019 Fast and high efficiency superconducting nanowire single-photon detector at 630 nm wavelength *Appl. Opt.* **58** 1868–72

[45] Borowski M and Leśniewicz M 2012 Modern usage of 'old' one-time pad *2012 Military Communications and Information Systems Conf. (MCC)*

[46] Mavroeidis V, Vishi K, Zych M D and Jøsang A 2018 The impact of quantum computing on present cryptography *Int. J. Adv. Comput. Sci. Appl.* **9**

[47] Gisin N, Ribordy G, Tittel W and Zbinden H 2002 Quantum cryptography *Rev. Mod. Phys.* **74** 145–95

[48] https://cryptosmith.com/2008/05/31/stream-reuse/ [Last accessed on 2021-08-23]

[49] Bennett C H and Brassard G 1984 Quantum cryptography: Public key distribution and coin tossing *Int. Conf. on Computers, Systems & Signal Processing* (New York: IEEE) pp 175–9

[50] Khanmohammadi A, Enne R, Hofbauer M and Zimmermanna H 2015 A monolithic silicon quantum random number generator based on measurement of photon detection time *IEEE Photonics J.* **7** 1–13

[51] Liao S-K *et al* 2017 Satellite-to-ground quantum key distribution *Nature* **549** 43–7

[52] Yin J *et al* 2017 Satellite-based entanglement distribution over 1200 kilometers *Science* **356** 1140–4

[53] Ekert A K 1991 Quantum cryptography based on Bell's theorem *Phys. Rev. Lett.* **67** 661–3

[54] Oberreiter L and Gerhardt I 2016 Light on a beam splitter: More randomness with single photons *Laser Photonics Rev.* **10** 108–15

[55] Khanmohammadi A, Enne R, Hofbauer M and Zimmermann H 2015 Monolithically integrated optical random pulse generator in high voltage CMOS technology *2015 45th European Solid State Device Research Conference (ESSDERC)* pp 138–41

[56] Bugl D 2017 Interface chip for random number generation in 350 nm CMOS *Master's Thesis* TU Wien DOI: 10.34726/hss.2017.35762

[57] Bude J, Sano N and Yoshii A 1992 Hot-carrier luminescence in Si *Phys. Rev.* B **45** 5848–56

[58] https://csrc.nist.gov/projects/random-bit-generation/documentation-and-software [Last accessed on 2021-08-23]

[59] John Walker. https://www.fourmilab.ch/random/ [Last accessed on 2021-08-23].

[60] George Marsaglia. https://web.archive.org/web/20160125103112/http://stat.fsu.edu/pub/diehard/ [Last accessed on 2021-08-23].

[61] Bisadi Z, Meneghetti A, Fontana G, Pucker G, Bettotti P and Pavesi L 2015 Quantum random number generator based on silicon nanocrystals LED *Proc. SPIE* **9520** 952004
[62] Dervić A, Tadić N, Mahmoudi H, Goll B, Hofbauer M and Zimmermann H 2020 Single-pixel postprocessing-free 5 Mbps quantum random number generator using a single-photon avalanche diode detector and a T/(T−t) pulse-shaped laser driver *Optical Eng.* **59** 1–10
[63] Fan-Yuan G-J, Wang C, Wang S, Yin Z-Q, Liu H, Chen W, He D-Y, Han Z-F and Guo G-C Dec 2018 Afterpulse analysis for quantum key distribution *Phys. Rev. Appl.* **10** 064032
[64] Aspuru-Guzik A and Walther P 2012 Photonic quantum simulators *Nat. Phys.* **8** 285–91
[65] Navarrete-Benlloch C, Pérez A and Roldán E 2007 Nonlinear optical galton board *Phys. Rev.* A **75** 062333
[66] Chandrashekar C M, Srikanth R and Banerjee S Aug 2007 Symmetries and noise in quantum walk *Phys. Rev.* A **76** 022316
[67] Broome M A, Fedrizzi A, Lanyon B P, Kassal I, Aspuru-Guzik A and White A G 2010 Discrete single-photon quantum walks with tunable decoherence *Frontiers in Optics 2010/ Laser Science XXVI* **104** 153602 Optical Society of America
[68] Crespi A, Osellame R, Ramponi R, Giovannetti V, Fazio R, Sansoni L, De Nicola F, Sciarrino F and Mataloni P 2013 Anderson localization of entangled photons in an integrated quantum walk *Nat. Photon.* **7** 322–8
[69] Sansoni L, Sciarrino F, Vallone G, Mataloni P, Crespi A, Ramponi R and Osellame R Jan 2012 Two-particle bosonic-fermionic quantum walk via integrated photonics *Phys. Rev. Lett.* **108** 010502
[70] Singh S, Chawla P, Sarkar A and Chandrashekar C M 2021 universal quantum computing using single-particle discrete-time quantum walk *Sci. Rep.* **11** 11551
[71] Wang H *et al* 2017 High-efficiency multiphoton boson sampling *Nat. Photon.* **11** 361–5
[72] Marvin M and Minc H 1965 Permanents *Am. Math. Mon.* **72** 577–91
[73] Arrazola J M *et al* 2021 Quantum circuits with many photons on a programmable nanophotonic chip *Nature* **591** 54–60
[74] Hofbauer M, Steindl B and Zimmermann H 2018 Temperature dependence of dark count rate and after pulsing of a single-photon avalanche diode with an integrated active quenching circuit in 0.35 μm CMOS *J. Sensors* **2018** 9585931
[75] Fukuda D, Fujii G, Yoshizawa A, Tsuchida H, Rathnayaka T, Takahashi H, Inoue S and Ohkubo M 2008 High speed photon number resolving detector with titanium transition edge sensor *J. Low Temp. Phys.* **151** 100–5 04
[76] Arute F *et al* 2019 Quantum supremacy using a programmable superconducting processor *Nature* **574** 505–10
[77] DiVincenzo D P 2000 The physical implementation of quantum computation *Prog. Phys.* **48** 771–83
[78] Scherer W 2019 *Mathematics of Quantum Computing* 1st edn (Berlin: Springer) https://doi.org/10.1007/978-3-030-12358-1
[79] Magnitskiy S, Frolovtsev D, Firsov V, Gostev P, Protsenko I and Saygin M 2015 A SPDC-based source of entangled photons and its characterization *J. Russ. Laser Res.* **36** 618–29
[80] Pittman T B, Jacobs B C and Franson J D 2001 Probabilistic quantum logic operations using polarizing beam splitters *Phys. Rev.* A **64** 062311
[81] Gasparoni S, Pan J-W, Walther P, Rudolph T and Zeilinger A 2004 Realization of a photonic controlled-not gate sufficient for quantum computation *Phys. Rev. Lett.* **93** 020504

[82] Koashi M, Yamamoto T and Imoto N 2001 Probabilistic manipulation of entangled photons *Phys. Rev.* A **63** 030301
[83] Walther P, Resch K J, Rudolph T, Schenck E, Weinfurter H, Vedral V, Aspelmeyer M and Zeilinger A 2005 Experimental one-way quantum computing *Nature* **434** 169–76
[84] Morris P A, Aspden R S, Bell J E C, Boyd R W and Padgett M J 2015 Imaging with a small number of photons *Nat. Commun.* **6** 618–29
[85] Erkmen B I and Shapiro J H 2008 Unified theory of ghost imaging with Gaussian-state light *Phys. Rev.* A **77** 043809
[86] Aspden R S *et al* Dec 2015 Photon-sparse microscopy: visible light imaging using infrared illumination *Optica* **2** 1049–52
[87] Moreau P-A, Toninelli E, Gregory T and Padgett M J 2018 Ghost imaging using optical correlations *Laser Photonics Rev.* **12** 1700143
[88] Duarte M F, Davenport M A, Takhar D, Laska J N, Sun T, Kelly K F and Baraniuk R G 2008 Single-pixel imaging via compressive sampling *IEEE Signal Process. Mag.* **25** 83–91
[89] Shapiro J H Dec 2008 Computational ghost imaging *Phys. Rev.* A **78** 061802
[90] https://andor.oxinst.com/learning/view/article/intensified-ccd-cameras [Last accessed on 2021-09-28]
[91] Sheppard C J R 2017 Resolution and super-resolution *Microscopy Research and Technique* **80** 590–8
[92] Khater I M, Nabi I R and Hamarneh G 2020 A review of super-resolution single-molecule localization microscopy cluster analysis and quantification methods *Patterns* **1** 100038
[93] Allen J R, Ross S T and Davidson M W 2013 Single molecule localization microscopy for superresolution *J. Opt.* **15** 094001
[94] Marc A 2018 Quantum-Optically Enhanced STORM (QUEST) for Multi-Emitter Localization *Sci. Rep.* **8** 7829
[95] https://andor.oxinst.com/learning/view/article/ccd,-emccd-and-iccd-comparisons. [Last accessed on 2021-09-28]
[96] Edamatsu K, Shimizu R and Itoh T 2002 Measurement of the photonic de broglie wavelength of entangled photon pairs generated by spontaneous parametric down-conversion *Phys. Rev. Lett.* **89** 213601
[97] Walther P, Pan J-W, Aspelmeyer M, Ursin R, Gasparoni S and Zeilinger A 2004 De Broglie wavelength of a non-local four-photon state *Nature* **429** 158–61
[98] Gasparini L, Bessire B, Unternährer M, Stefanov A, Boiko D, Perenzoni M and Stoppa D 2017 SUPERTWIN: towards 100 kpixel CMOS quantum image sensors for quantum optics applications *Proc. SPIE* **10111** 101112L
[99] Zhong H-S *et al* Dec 2018 12-Photon entanglement and scalable scattershot boson sampling with optimal entangled-photon pairs from parametric down-conversion *Phys. Rev. Lett.* **121** 250505
[100] Mankei T 2009 Quantum imaging beyond the diffraction limit by optical centroid measurements *Phys. Rev. Lett.* **102** 253601
[101] Toninelli E, Moreau P-A, Gregory T, Mihalyi A, Edgar M, Radwell N and Padgett M 2019 Resolution-enhanced quantum imaging by centroid estimation of biphotons *Optica* **6** 347–53

www.ingramcontent.com/pod-product-compliance
Ingram Content Group UK Ltd.
Pitfield, Milton Keynes, MK11 3LW, UK
UKHW051854100926
8556IPUK00005B/615

9 780750 325820